Progressions and Directions

Charles A. Jayne

ISBN-10: 0-86690-615-0
ISBN-13: 978-0-86690-615-9

Cover Design: Jack Cipolla

Published by:
American Federation of Astrologers, Inc.
6535 S. Rural Road
Tempe AZ 85283

www.astrologers.com

Printed in the United States of America

Dedicated to

Edgar R. Wagner, Ph.D.

Contents

Foreword

The first part of this book describes progressions: secondary or major, minor, and tertiary. The second part is a full description of certain new directions in astrology.

The third part is a discussion of the knotty problem of the valuable but neglected primary directions. First the usual ones are described, including directions under a body's own polar elevation, the latter improperly calculated by Zadkiel, Sepharial, DeLuce, Kuehr and others, as Fagan noted. Mention is made of primaries *in zodiaco* under the body's polar elevation which results in the vertical and horizontal longitude positions. Then a description is given of the primaries used so successfully by Johndro. No attempt is made to deal with these various techniques at length except for the new directions, which are primaries *in zodiaco* sine (celestial) latitude. They are easy to do when compared with other forms of primary direction.

A fourth section focuses on symbolic directions. There is one useful tool out of a number of invalid modes of symbolic direction.

This publication covers all forms of progression and direction in an attempt to point out the valid ones. There is unfortunately no book that adequately deals with all methods.

A fifth section gives examples of what has been discussed with special reference to the little understood primary directions. By following these examples any student should be able to use all of the methods described. In the last part of the section on the primary directions are two pages of diagrams. Very often a visual example of something makes it a lot clearer. Indeed, it would be advisable to draw small rough diagrams of the celestial sphere for yourself as a crosscheck on your calculations. Too few astrologers actually visualize the space about them in which the celestial bodies live and move and have their being!

Double interpolation, which is often required to find the exact degrees and minutes of arc on an Ascendant from an accurate Midheaven, is explained in the Appendix. Triple interpolation is also explained, but merely means carrying double interpolation one step further.

The Appendix also has a table for finding right ascension from the longitude and latitude (celestial). This makes it possible to work those primary directions that involve aspects to or by the Midheaven.

Chapter 1

Progressions

Let us consider the astronomical basis of both progressions and directions. Earth has three fundamental motions, not just two. It rotates on its axis once a day along the plane of the equator. The Earth and Moon together revolve about the Sun in one year along the plane of the ecliptic. These two motions are well known, as are the planes of the equator and ecliptic. Indeed, the tropical zodiac commences at the North Node of the ecliptic on the equator, i.e. at 0°00′ Aries, where the ecliptic crosses above the equator. Celestial longitude is measured along the ecliptic plane from 0° Aries, as is right ascension. But the latter is measured along the equator.

In astrology, positions *in zodiaco* are given in celestial longitude as a rule, and some arcs of direction are longitudinal. But arcs in right ascension are just as valid and as important, as we shall see later.

The Moon does *not* revolve around Earth! Staying always exactly opposite to each other, they revolve around their common center of mass, or barycenter, in 27.32 days, this being the third fundamental motion of Earth that takes place along the plane of the Moon's orbit. This lunar plane is tilted (mean) 5°09′ to the solar plane (ecliptic). The place where the Moon's orbit plane crosses above the ecliptic is the Moon's North Node. The center of the Earth is 2,922 miles (mean) from the Earth-Moon barycenter, and so Earth describes a very small elliptical orbit in space in 27.32 days. Thus we have three periods of fundamental motion: the day, the lunar period, and the solar period (year).

In progression we say that a longer one of the three periods is equivalent in life to the motions of the lights and planets during a shorter one of the three periods. If one day in the ephemeris is equivalent to one year of life we have the secondary, or major, progression. If one lunar period is considered to be equivalent to one solar one (27.32 days = 365.25 days) we have the minor progression (a ratio of 13.368). Edward Troinski of Germany discovered that if we allow a day to be equivalent to one lunar period (ratio is 27.32) we have his tertiary progression. This last form of progression is the newest and has been tested by a number of capable astrologers who find it to be valid and useful.

By progression all bodies move at their own and different rates. They may move direct or retrograde. By direction they move at the same or nearly the same speed and must all move in the same direction. In the case of the secondary progression the ratio is 365.25 (Fagan and Johndro use a ratio of 1 to 366.25 instead of 1 to 365.25). Therefore, the planets move more slowly by this progression than by either the minor or tertiary methods, thus making secondary progressions more potent. As the minor progression has the smallest ratio

(1:13.368), its rates are the fastest and the least potent of the three. The faster a body moves the less powerful its effects but the more accurate its timing.

The secondary Moon moves between 12° and 15° a year so that its effects are much more minor than most astrologers appear to realize. While the planets beyond Mars move very slowly and thus very potently by this method, they make aspects so seldom in life that their usefulness is limited. Only the Sun, Mercury, Venus, and Mars move fast enough to make fairly frequent aspects, yet slow enough that they are not minor. A secondary Mars is potent, but the event can considerably lead the exact time of the aspect. By tertiary progression the planets inside the orbit of Jupiter move fairly fast and so are rather minor. Only Jupiter and Saturn move at a middle speed, and Uranus, Neptune, and Pluto are still fairly slow. Nearly the same thing is true of the minor progression. All three methods are valid and useful but have the limitations described. They take no account of either celestial or terrestrial latitude.

Three years in the ephemeris are equivalent to about 40 years of life by minor progression. One year in the ephemeris is equivalent to 27.33 years of life by tertiary progression so that 1.5 years are equivalent to nearly 41 years. One can put it this way: each lunar return is equivalent by minor progression to a solar return and thus a year of life. By the slower tertiary progression each diurnal return is equivalent to a lunar return. But bodies may be "regressed" as well as "progressed." That is to say that by secondary progression each day *before* birth is also equivalent to one year of life. I term such retrograde motions *regressions*. They do work but are not as potent as the *progressions*.

Whether dealing with progressions or regressions one should always take account of bodies having the same declination, i.e. parallels, or those having opposite declinations (one north and the other south), which are termed contraparallels and act like oppositions. Such aspects can be quite important and should always be included. The matter of how to progress, or direct, the Midheaven and Ascendant are addressed in Chapter 2.

Secondary progressions are much easier to calculate if one uses what has been termed the Adjusted Calculation Date (ACD), and is also referred to as the Limiting Date or Artificial Birthday. It is assumed that one is using an ephemeris based on either noon or midnight Greenwich time (GMT). It is also assumed that one has a fairly accurate time of birth. Now one day equals one year so that 24 hours must equal 12 months. Therefore we divide both sides by 12 such that each two hours in the ephemeris must then equal one month of life. Since Greenwich noon is eight hours before an 8:00 pm birth time, it is equivalent to 8/2 = 4 months *before* birth by progression, thus making the ACD four months before the birthday. Thus if birth had occurred November 10, the ACD would be four months earlier, or July 10. Then 30 *days* later in the ephemeris would be equivalent by progression (secondary only) to July 10, which is 30 *years* later (and *not* on the 30th birthday in November). Since the planets are given for noon, this requires no calculation to find out where they are.

Let us say that instead we use a midnight ephemeris. We want the number of hours that birth occurred since the prior midnight as this is a midnight ephemeris, so we must *add* 12 hours to the 8, giving us 20 hours after midnight. Then 20 divided by 2 = 10 months. Since birth occurred *after* midnight, we must go backward to reach it and thus must also go backward 10 months to reach the ACD, which occurred January 10.

Let us say that the GMT of birth had been at 6:00 am and we were using a midnight ephemeris. Then 6 divided by 2 = 3 months, which again must be *subtracted* from the birthday in order to obtain the ACD. But if the ephemeris were a noon one, then noon is 6 hours *later* than the time of birth, so 6 divided by 2 = 3 months, which is now *added* to the birth date since the time is *later*.

Now we said that two hours are equivalent to one month, so that 120 minutes of time are equivalent by progression to 30 days of life. If we then divide both sides by 30, we find that each 4 minutes of time equal one day. A man born at 2:24 pm GMT and using a noon ephemeris gives: 2 hours divided by 2 = minus 1 month and 24 minutes divided by 4 = minus 6 days. Thus we *subtract* 1 month and 6 days from his birthday to find his ACD. If we a are *regressing* the chart we should do the opposite, i.e., we would *add* the 1 month and 6 days to find the converse ACD. Thus it is easy to find the converse ACD for regressions since it always is the same interval from the birthday as the ACD, but in the other direction.

Chapter 2

New Directions in Astrology

Many astrologers would agree that the most important tool that an astrologer can use in the dynamic or timing phase of astrology is a method of direction or progression. At the present time most astrologers accept and use secondary, or major, progressions in which the positions of the planets so many days after birth are thought to be equivalent to their positions by progression the same number of years after birth in the life of the individual whose chart is so progressed. The experience of our own research group is in accord with this general consensus.

We believe that we are qualified to speak with some assurance about these methods because of the amount of experience we have had in their use and their testing. E. Richard Wagner commenced using solar arc directions forty-five years ago at about the time that Alfred Witte commenced to use the method in Germany. As Sieggruen pointed out, this method of direction has been rather widely used in Europe and has therefore been tried and tested by other individuals and groups besides our own. As the length of the arc is not affected by an incorrect birth time, and since it is extremely easy to determine, it is the most valuable method of direction known to us. Since it is also based on the *true* motion of the Sun, and not on the fictitious mean motion—the measure of Naibod, or 0°59′08.33″—or the equally fictitious one degree a year of Ptolemy, it has a sound astronomical basis in fact. More will be written later about the astronomical basis for all three methods of direction.

In the spring and fall the true motion of the Sun for a time is the same as this mean (Naibod) motion, which is one reason why so many astrologers have been unable to determine which "key" is the right one. In a similar manner at two different times of the year the true motion of the Sun equals one degree, and thus this is one reason why some astrologers have been unable to choose between these two methods. But another and vital reason for their difficulties has been the ridiculously large orbs of aspect (as much as plus or minus one degree, i.e., an uncertainty of plus or minus one year!) that they have permitted in their work. Some otherwise very capable and experienced astrologers have made this error. For when considerably narrower orbs are employed, one finds that the Naibod and one degree "keys" simply do not give accurate or reliable results. We are sure of this since five of us have worked on about 2,500 charts.

The orbs we use seldom exceed minus 5′ (event before direction by one month) and plus 15′ (event lags after direction by three months). We have found that at least two-thirds of the events occur within minus 1.25′ and plus 3.75′, i.e., a total range of one month, and another twenty percent within minus 2.5′ and plus 7.5′, i.e., a total range of two months. In professional work any range of uncertainty in excess of such values would be of little use. The planet whose influence by direction is most likely to lead the exact date of the direction is Neptune (and secondary progressed Mars). On the other hand, Saturn's effects nearly always lag the date of the direction and this is also true to a lesser degree of the Sun. The most accurate timer is certainly Uranus.

In the radix method of direction of the late Vivian Robson all the planets and the Moon were moved in longitude the same distance that the Sun moved by its mean (Naibod) motion. The solar arc method is the same except for the vital difference that all bodies are moved the same distance as the Sun in its *true* motion. Solar arc is a radix method insofar as all bodies maintain, in their directed positions, the same relationship with each other that they had in the radix. In early July when Earth is at its aphelion, the Sun's apparent motion is 0°57′12″, or nearly 2′ a day (a year by direction) less than its mean or Naibod motion. Thus a person born in June has a solar arc that is nearly 10 less than the Naibod arc at age 30, showing how large and how important the divergence between the two arcs may be. When Earth is at its perihelion in early January the Sun moves 1°01′09″ a day (61′09″).

Solar Arc in Longitude

To find the solar arc one measures the distance or arc that the Sun has moved by secondary (major) progression. This may be done in either of two ways. If one has an accurate time of birth then one may use the ACD (also termed Limiting Date and Artificial Birthday). Then the arc from the natal Sun to the progressed Sun, for any given year, month and day of life, is the length of the solar arc. Or one may find the arc, if the natal time is unknown or inaccurate, by using the place of the

Sun in the ephemeris for the day of birth as the base position from which to measure (rather than its position in the chart). Then the solar arc at age 42 years and 3 months will be the position of the Sun in the ephemeris 42 days after birth less its position in the ephemeris for the day of birth—giving the solar arc for 42 years. To find the extra motion in three months, since this is one-quarter of one year and since a direction year is a day in the ephemeris, one multiplies the Sun's motion in one day by one-quarter. The Sun moves approximately 5′ in one month by direction. Thus we see that it is very easy indeed to find the length of the solar arc.

Assuming that the solar arc is direct (not converse, or prenatal) it is then added to every planet in the chart and to the Moon and Midheaven. Thus even if Mars may be retrograde at birth, it is caused to move direct by this method of direction since the Sun's motion is direct. And you will note that the solar arc is used to direct the Midheaven. Indeed, this motion of the Midheaven, as will later be explained, is the real basis for this whole solar arc method of direction. We cannot possibly recommend this method of direction too strongly for it is simple, accurate, reliable, potent, and unaffected by inaccuracy in the birth time. The Moon by this method moves roughly 1° a year as contrasted with the 12° to 15° annual motion in the secondary progression and is therefore far more potent.

Ascendant Arc

One of my main associates in research, Eleanor Hesseltine, discovered the Ascendant arc method of direction. When one has found the position of the directed Midheaven by adding the solar arc to the natal Midheaven, one is then able to find an Ascendant in a tables of houses that corresponds to this new or directed Midheaven, using always the terrestrial latitude of the natal chart and not that of the locality at the time of the direction. (See Appendix for a description of the double interpolation method required to find an Ascendant. This method can be used for any required double interpolation, so important in astrology.) Then the distance or arc from the natal Ascendant to this directed Ascendant is the Ascendant arc. This arc may then be added to all the natal planets and the Sun and Moon quite independently of their solar arc positions. Thus anyone who can accurately find the Ascendant and who has an accurate (rectified) chart can find the Ascendant arc with ease.

One may never direct the Midheaven by Ascendant arc or the Ascendant by solar arc. If the birth time is erroneous the natal Midheaven will be inaccurate and so the position of the directed Midheaven will be equally inaccurate, as will the natal and directed Ascendants. But, in addition, the *length* of the Ascendant arc will also be erroneous, the degree of inaccuracy in its length depending partly on the amount of inaccuracy in the birth time and partly on what part of the zodiac the Ascendant is in. For if the Ascendant is in a sign of short ascension where it moves rapidly—Aquarius, Pisces, Aries, Taurus—the error in the length of the Ascendant arc will be greater. Thus it is advisable not to use Ascendant arc directions unless the time of birth has been carefully rectified. If the error in the time is small and if the natal Ascendant is in a sign of long ascension—Cancer, Leo, Virgo, Libra, Scorpio, Sagittarius—the Ascendant arc will be fairly accurate.

Unfortunately the great majority of astrologers do not use the correct terrestrial latitude. There are three latitudes on Earth, i.e., distance in arc that a place is above or below the equator. The astronomical latitude (gravitationally determined) differs but minutely from the ordinary geographical latitude, which is found on all maps and is usually, and erroneously, used by many astrologers.

The geocentric latitude is the correct one to use in astronomy and astrology and is always less than the ordinary geographic latitude. This geocentric latitude is the angle between the equator and a radius from the center of Earth to the surface where the person is located. In the geographic latitude one drops a perpendicular to the plane of the horizon at the locality and then measures the angle that this perpendicular makes with the equator. Only at the poles of Earth or at the equator does this perpendicular pass through the center of Earth such that the two types of latitude are then identical. The discrepancy between these two latitudes is due to the fact that Earth is not perfectly spherical; it is an oblate spheroid.

Below is the table of corrections to convert ordinary or geographic latitude into geocentric latitude, the correction always being a subtraction whose value ranges from 0′ at the poles and equator to 12′ at 45° north or south geographic latitude (actually the maximum value is 11′35.65″). Values in the table are to the nearest minute of arc, and for degrees of geographic latitude are to the nearest degree. In order to accurately determine either a radical or a directed Ascendant, or Vertex, the geocentric latitude must be used.

Between values of latitude given on the left below (values are inclusive), use the corrections to geographic latitude on the right to find geocentric latitude.

0-1°, 89-90°	-00′
2-3°, 87-88°	-01′
4-6°, 84-86°	-02′

7-8°, 82-83°	-03′
9-11°, 79-81°	-04′
12-14°, 76-78°	-05′
15-17°, 73-75°	-06′
18-20°, 70-72°	-07′
21-23°, 67-69°	-08′
24-27°, 63-66°	-09′
28-32°, 58-62°	-10′
33-41°, 49-57°	-11′
42-48°	-12′

With the Ascendant near the cusp of Aquarius or Gemini at London, the error in the Ascendant (the difference between geographic and geocentric latitude) is slightly over 17′. By a slow direction this can produce, therefore, an error of more than four months in its timing. Such an error is unacceptable in any serious or professional work.

Vertex and Vertical Arc

Since space has three dimensions, not two, a horoscope is also three dimensional and is not limited to the two dimensions of the sheets of paper on which we draw it. For this reason a horoscope has three angles and not merely the two we are accustomed to using. The intersection of the ecliptic with the meridian produces the Midheaven-IC axis. The intersection of the ecliptic plane with the plane of the horizon produces the Ascendant-Descendant axis.

But any locality has three local planes perpendicular to each other and not just two. In addition to the meridian (perpendicular to Earth and running north and south like the north-south wall of a room) and in addition to the horizon (a horizontal plane like the floor of the room) is the prime vertical (perpendicular to Earth and the other two planes and running east and west like the east-west wall of a room). The intersection of the plane of the ecliptic with this third local plane (prime vertical) creates a third axis and thus a third angle for every horoscope. As this axis lies in the prime vertical, I term its west end the Vertex and its east end the Anti-vertex. Indeed, only occasionally is the Ascendant due east and the Descendent due west, but the Vertex is always due west and the Anti-vertex is always due east. When the Sun is due west of your locality, it is conjunct the Vertex even though it is below the horizon part of the time and above it at other times.

At first I thought that the east end of this axis was the significant one, but directing by it produced no results, whereas true direction of the (western) Vertex did produce results. Johndro did much work with this angle, terming the Anti-vertex the electrical Ascendant (and the conventional Ascendant, the magnetic Ascendant). But in a letter to me shortly before his death he concurred with me that by oblique ascension the western or Vertex end was the significant one. I am more concerned here with the use of the directed Vertex to obtain the third or vertical arc than I am with its static meaning and function. First, let us consider how to find the Vertex using an ordinary tables of houses. (For those born between 30° north and 30° south it may only be found by trigonometric calculations as is the case with the Ascendant for latitudes between 60° north or south and the poles.)

To find the Vertex, radical or directed, one uses the geocentric co-latitude instead of the geocentric latitude. The co-latitude is the distance the locality is from the pole instead of the equator. Therefore, the sum of the latitude and co-latitude is always 90°. To find the geocentric co-latitude, subtract the geocentric latitude from 90°. Thus, if the geocentric latitude is 40° north, the geocentric co-latitude is 50°, i.e., 90° - 40°. Just as we must use the Midheaven and the latitude to find the Ascendant, so to find the Vertex we must use the IC (cusp of the fourth house) and the co-latitude. If we enter the tables of houses using the IC in place of the Midheaven and the co-latitude in place of the latitude, the resultant rising degree is actually the Vertex. Thus the Vertex is not difficult to determine.

The distance from the natal or radical Vertex to the directed Vertex is the vertical arc. Using the co-latitude and the point opposite the directed Midheaven, i.e., the directed IC, we find the directed Vertex in any tables of houses (just as we would find an Ascendant using the directed Midheaven and the latitude). But again, if the natal time is inaccurate the length of the vertical arc will be erroneous, the amount of error depending on the time deviation from the true time and also on whether the Vertex is, like the Ascendant, in a sign of short or long ascension. A rapidly moving vertical or Ascendant arc (Ascendant or Vertex in a sign of long ascension) will be more potent than solar arc. Of course the Vertex may never be moved by solar or Ascendant arc but should be directed by vertical arc.

As a rule the Vertex lies in the fifth, sixth, seventh, or eighth houses. With the Midheaven at 0° Capricorn, the Ascendant at 0° Aries is exactly opposition the Vertex at 0° Libra; with the Midheaven at 0° Cancer, the Vertex at 0° Aries is exactly opposition the Ascendant at 0° Libra. Otherwise the Ascendant and Vertex are not opposition each other. The Vertex refers to the role or function of the individual and those factors in the life that are completely fated and from the past (karmic), and it relates the individual to the group and is the least personal of the three angles.

The Anti-vertex is the Ascendant of the horizontal house system. In the case of a premature child, whose gestation period is in the vicinity of seven months, the Vertex of the birth chart determines, by conjunction or opposition, the place of the Moon in the conception chart (and the Vertex of the conception chart in such cases determines the Moon in the natal chart). The significance of the Vertex has been much neglected in astrology. The vertical arc was discovered by me.

Thus each radical body, aside from the Sun, has four positions in longitude: by secondary progression, by solar arc, by Ascendant arc, and by vertical arc. If one also employs converse solar arc, then one has also converse Ascendant and vertical arcs. It has been our experience that these converse arcs are valid but are less potent than direct arcs. Arcs of the type we are describing, which I have used in extensive professional work, are very useful for testing hypothetical planets, for example, since it is seldom that the right ascensions or declinations are known, whereas an approximate value of their longitudes is usually known. Furthermore, the use of these three arcs enables one to predict most life events since by the secondary progression alone there are so few progressed aspects per year.

Let us give an example of the determination of all three arcs. Consider a native born with the Midheaven at 0° Capricorn, the Ascendant at 0° Aries and the Vertex at 0° Libra. Let us also assume that the geocentric latitude is 40° north and that the natal Sun is at 16°28′49″ Gemini (June 7, 1899), birth having occurred at Greenwich noon. Now we shall find the three arcs at age 30, or 30 days in the ephemeris after June 7,

1899. On July 7, 1899, the Sun was at 15°06′19″ Cancer. Then the solar arc will be 15°06′19″ Cancer less 16°28′49″ Gemini, or 28°37′30″, i.e., 28°37′30″. We now add this to the Midheaven, i.e., 0°00′ Capricorn plus 28°37′30″, which places the directed Midheaven at 28°37′30″ Capricorn. At 40° north geocentric latitude a 28° Capricorn Midheaven gives an Ascendant of 17°29′ Taurus, and a 29° Capricorn Midheaven gives the Ascendant at 18°54′ Taurus, a motion of 1°25′, or 85′ for the Ascendant, while the Midheaven moves 1° or 60′. But we wish the Midheaven to move only 37′30″ so we have (37′30″/60′) x 85′, which equals slightly over 53′. Adding this to the 17°29′ Taurus Ascendant we find that the directed Ascendant has reached 18°22′ Taurus. Then the Ascendant arc will be this position in Taurus less the 0°00′ Aries position of the natal Ascendant, an Ascendant arc of 48°22′, considerably more than the solar arc.

Now if the directed Midheaven is 28°37′30″ Capricorn, the directed IC is 28°37′30″ Cancer (same value in the opposite sign) and we use this as though it were a Midheaven and enter the tables of houses at a co-latitude of 50° since 90° minus 40° (geocentric latitude) equals 50° of geocentric co-latitude. The Vertex at this co-latitude for a “Midheaven" of 28° Cancer is 21°35′ Libra, and its value for a “Midheaven” of 29° Cancer is 22°19′ Libra. The difference between these two Vertices, 0°44′, is the distance the Vertex has moved while the “Midheaven” (IC, actually) has moved 60′. So we find the value of (37′30″/60′) x 44′ or 0°27′30″, which is added to the 21°35′ Libra Vertex to find the correct position of the directed Vertex, i.e., 22°02′30″ Libra. Then if we subtract the radical Vertex at 0°00′ Libra from this we find the value of the vertical arc or 22°02′30″, which is less than half of the Ascendant arc.

In order to make full use of these three arcs we need to know the annual rate of advance of each arc. In the case of the solar arc we look to see how far it moves between July 7, 1899 and July 8, 1899, i.e., in the next year by direction. This value is 16°03′33″ Cancer less 15°06′19″ Cancer, or 0°57′14″, which is the annual rate of the solar arc. If we divide this by 12—calling it 57′15″—we find the rate of the monthly arc, i.e., 4.77′. Now if the Midheaven moves 60′ we noted that the Ascendant would move 85′, where it now is in Taurus by direction, but the Midheaven is moving at the annual rate of the Sun—57′15″—so that the annual rate of the Ascendant is (57′15″/60′) 85′ or 81.1′. Dividing this by 12 we obtain a monthly rate of 6.76′. In the part of the zodiac where the directed Vertex now is, it moves 44′ for every 60′ the "Midheaven" moves, so its annual rate is (57′15″/60′) x 44′ or 42′, and the monthly rate is 42′/12 or 3.5′. Thus with comparative ease it is possible to find the annual and monthly rates of the three arcs.

If the annual rate is nearly constant, as it is in the signs of long ascension (Cancer, Leo, Virgo, Libra, Scorpio, Sagittarius), it may be used for several years without any recalculation. As the directed Vertex is in Libra, such is the case for it, but not for the directed Ascendant which is in a sign of short ascension (Aquarius, Pisces, Aries, Taurus). Both angles move at an intermediate speed in Gemini and Capricorn.

One is sometimes inclined to believe that the simpler something is, the harder it is to discover. Such is the case, for instance, with these three arcs and the striking astronomical basis that they have. To those who know primary arcs, the terms MD (meridian distance) and HD (horizontal distance) will be familiar. Only here we measure the distance of a planet from the meridian, and Midheaven, in longitude and along the ecliptic instead of in right ascension and along the equator. Solar arc directions are based on maintaining the meridian distance of a planet's constant, i.e., the planets are directed at the same rate, in longitude, that the Midheaven is advanced, and keep a constant distance from it. If we measure the distance from the horizon (HD) and Ascendant in longitude, then the horizontal distance is kept constant in the Ascendant arc, i.e., it maintains a constant distance from the Ascendant and moves at its rate. The VD, or vertical distance, in longitude, from the prime vertical and Vertex is kept constant in the vertical arc form of direction.

It is generally agreed that for any major progression or direction, the ratio of one solar day is equal to one tropical year (1:365.2422) will be used (although Johndro and Fagan advocate one sidereal day as equal to one sidereal year—1:366.25 ratio). Now at true local noon the Sun is conjunct the Midheaven, i.e., is crossing the upper meridian. And between two such successive true local noons—or one true solar day—the Sun advances from 57′ to 61′ (depending on the time of year), but by direction this is its motion in one year. Now as the Midheaven is conjunct the Sun, by definition *it too must move forward by the same arc.* Nor does this have anything at all to do with the equator or the right ascension of the Sun; indeed, we are dealing with the motion of the Sun on its own plane, i.e., the ecliptic, the major plane of reference for astrology where all aspects are usually read.

Thus the whole system we are advocating is based on this simple astronomical fact that the daily motion of the Sun creates an equal motion of the Midheaven in longitude, and that by direction this becomes the annual motion, the true motion of the Sun and not the fictitious mean motion or the even more fictitious one degree value.

But for every advance of the Midheaven there is a related, but seldom equal, advance of the Ascendant. Indeed, the Ascendant is caused, thereby, to advance by a different rate and thus to create an arc of a different length. The same line of reasoning applies also to the Vertex and its motion. Indeed the motions of the Ascendant and Vertex are determined by the motion of the Sun and Midheaven, and by the latitude and co-latitude (geocentric) of the locality. Just as a prism refracts white light into the three primary colors, so the three local planes—representing the three dimensions of space—"refract" the single motion of Earth into three components that are the three arcs we have described.

And we could use the three primary colors for them: blue-violet for the Midheaven, red for the Ascendant, and green for the Vertex. One might even surmise that these may be the real primary directions, for certainly they are just as primary for directional astrology as the three primary colors are for the whole domain of light.

But the ultimate test is the pragmatic one: How well do they work when put to the test? I earnestly ask you to test these methods. At least do not reject them until and unless you have tested them carefully. In our experience they are astrology's most valuable tool.

Directions in Declination

We now come to directions in declination. Some astrologers have raised their eyebrows at the use of declination in astrology because it belongs with the equatorial system of right ascension and declination, whereas zodiacal longitude belongs with the ecliptic system, i.e., of celestial longitude and celestial latitude. But the pragmatic experience of our ABC Research Group (Astrological Bureau of Consultation), and of many other astrologers, ancient as well as modern, supports the validity of the use of declination in astrology. Declination is the distance in arc that a body is from the equator, north or south of it, along a great circle which cuts the equator at a right angle and passes through both the north and south poles of Earth.

In astrology it has long been customary to consider that two bodies which have the same or nearly the same declination north (or the same declination south) are in parallel, and act as though they were in conjunction. But a parallel is more psychological and less tangible than a conjunction is in its influence.

Many astrologers consider that if two bodies are the same, or nearly the same, distance from the equator, one north and the other south, that they, too, are in parallel. But we have found that they act as though were in opposition rather than in conjunction, and call this aspect the contraparallel.

We also find that the usual conception of orbs is mistaken in connection with these parallels and contraparallels. For two bodies a certain distance apart in declination near the equator, i.e., having *low* declinations, are effectively much farther apart when far from the equator, i.e., having *high* declinations, even though they may be the same distance apart in declination. A change in declination of 24' close to the equator is equivalent to about 1° of longitude, whereas a change in declination, of the Sun for instance, in high declination (the Sun never exceeds 23°27′ north or south) of only 1′ can be equivalent to 1° of longitude.

Therefore, in measuring the orb of declination aspects we suggest that you should transform any given declination, provided it is less than 23°27′ north or south, into its equivalent position on the ecliptic. This is easy to do. Let us say that a planet has a certain value of north declination. Then find the longitude that the Sun has (from an ephemeris) when it has the same declination as the planet. Then this longitude you have found is the ecliptic-equivalent. Actually there will always be two of them such that they are equidistant from, and on opposite sides of 0° Cancer or 0° Capricorn.

Thus the point on the ecliptic of 20° Gemini—10° back from 0° Cancer—has the same declination as the point on the ecliptic at 10° Cancer–10° forward from 0° Cancer. Once one has found all the ecliptic equivalents of the Sun, Moon, planets, and the three angles (Midheaven, Ascendant, and Vertex) one may say that any two bodies are in parallel if the distance apart of their ecliptic equivalents is less than the largest orb you would use for the conjunction of the same two bodies. Or, if they are, via their ecliptic equivalents, close enough to an opposition in longitude, then actually they are in contraparallel.

What has been said thus far cannot, however, be applied to those bodies above what we call the "turn," i.e., 23°27′ north or south, where the Sun turns from increasing to decreasing declination. These are the solstices (0° Cancer and 0° Capricorn).

Solar Declination Arc

Just as we may take the arc or distance that the Sun has moved in longitude along the ecliptic, so we make the distance that the Sun had moved in declination from its natal or radical position in declination. The result is termed “solar declination arc,” which can be used to direct the Moon, Midheaven, and all the other planets.

Let us say that at birth the Sun is 2°21′ north declination and that at a later date it has moved to 12°26′ north declination. The difference between these two values—10°05′—is the solar declination arc. If the Sun had been 2°21′ south declination at birth and had moved to 12°26′ north declination, we would have to add the two values since the Sun would have first moved from below the equator (south) to the equator (when it crosses the equator the declination is 0°00′) or 2°21′, and then it would have moved all the way to 12°26′ north or an additional 12°26′ and a total solar declination arc of 14°47′.

It is advisable to indicate on the natal chart whether a given body is increasing or decreasing in declination. For instance, suppose that in the last example, Pluto at birth is at 10°00′ north declination. If it is decreasing at birth, which means it is moving closer to the equator so that the numerical value of the declination (north or south) is growing less, then we must subtract the solar declination arc from the natal declination. It will take 10°00′ of the solar declination arc of 14°47′ to reach the equator. If we subtract this 10°00′ from the 14°47′, we find how far Pluto is directed on the other side of the equator, in southern declination (at birth it was in northern declination). There it would reach 4°47′ south declination by direction and would, by further direction, be moving into ever higher south declination. If in so doing it reached the same declination as another planet, then they would be in parallel, described as: Pluto parallel X by solar declination arc. As this arc can be very slow, its effects can be powerful.

But let us say that at birth, or radically, Pluto was moving north, i.e., was increasing in declination. Then 14°47′ added to 10°00′ north becomes 24°47′ north declination by direction. However, this would take it beyond the turn at 23°27’ (the highest declination of the Sun) so we have to find out how far it would move past 23°27′. 23°27′ subtracted from 24°47′ is 1°20′. Since by solar declination arc no body may go past the turn (this is important) we must bring Pluto back 1°20′ from the turn, i.e., 23°27′ minus 1°20′ or 22°07′, which thereby becomes the position of Pluto in declination (still north) by solar declination arc. Only now it is decreasing in declination, in its further motion by direction, whereas at birth, or radically, it was increasing. We make all planets turn from increasing declination to decreasing declination at the same value of maximum declination as that of the Sun since the Sun turns there, too, and since this is solar declination arc.

This same principle must be kept in mind in finding the value of the solar declination arc in certain other instances. Let us say that at birth the Sun has a declination of 22°50′ south, i.e., is in Sagittarius, and the declination is increasing. At a later date the declination has become, say, 20°12′ south but is now decreasing. First it moved from 22°50′ south to 23°27′ south (to the turn)—a motion of 0°37′. Then it moved from 23°27′ south to 20°13′ south (now decreasing in declination), a motion of 3°14′. Then in its total motion, the solar declination arc will be 0°37′ plus 3°14′ or a total 3°51′. If you study an ephemeris, all of this should present no problem.

Let us say that having found this 3°51′ arc we wish to direct Mars by solar declination arc, and that Mars at birth is 24°56′ north and is increasing in declination, i.e., going still higher in declination even though it is already above the turn. In spite of this we cause it to decrease, i.e., bring it down from 24°56′ north to 21°05′ north. For had we been progressing Mars by secondary (major) or tertiary (minor) progression, we would have allowed it to continue to move further north in declination since that is its actual motion. Indeed, it might not have turned to decreasing declination until it reached over 25°, but since the method of declination arc we are using is that of the Sun, it must in all such cases be turned immediately to decreasing declination.

All that is written here will also apply to direction by both Ascendant declination arc and vertical declination arc since the directed Ascendant or Vertex may never have a value of declination greater than the maximum declination of the Sun itself. Solar and Ascendant declination arc directions were discovered and tested by Eleanor Hesseltine and E. Richard Wagner. I discovered vertical declination arc.

Ascendant Declination Arc

It is easy to find the value of the Ascendant declination arc. One first must know the declination of the natal Ascendant. The declination of the Ascendant will always equal the declination of the Sun when it has exactly the same longitude as the Ascendant. So one looks in an ephemeris at the value of the Sun's declination when it has the same longitude as the Ascendant. Then one finds the declination of the directed Ascendant using exactly the same method as for finding the declination of the natal Ascendant. Then the distance of arc that the Ascendant has moved in declination from the natal value to the directed value is the Ascendant declination arc. If the Ascendant moves from Gemini—where, like the Sun, it has a north declination and is increasing in declination—to Leo, where the Ascendant, while still north is now decreasing in declination, one measures the distance in declination from the natal position in Gemini to the turn at 0°00′ Cancer—where it is 23°27′ north—and then the distance in declination again, from the turn (23°27′ north) to the value of Leo. Then the sum of these two distances will be the Ascendant declination arc.

Of course if the chart has an inaccurate Ascendant (and Vertex), i.e., has not been rectified, then the length of the Ascendant (and vertical) declination arc, will be erroneous. The error may be quite small if the natal Ascendant is itself only in error by a small amount. In general the error will be greater if the natal or the directed Ascendant is in Aquarius, Pisces, Aries or Taurus, and will be less in signs from Cancer through Sagittarius. In the southern hemisphere this will be true in the opposite signs.

Vertical Declination Arc

One finds the declination of the natal Vertex just as one finds that of the Ascendant, and the declination of the directed Vertex just as one finds that of the directed Ascendant. Then the motion in declination from the natal Vertex to the directed Vertex gives the vertical declination arc just as we find and measure the Ascendant declination arc.

If a planet is at 22°30′ north and is decreasing, and another one is at 12°15′ north and is also decreasing, a declination arc (solar, Ascendant, or vertical) of 10°15′ will bring the planet in the higher declination into parallel with the one in lower declination. If, however, the one in the lower declination is increasing instead of decreasing in declination at birth, then at the same time that the planet of higher declination is brought to a parallel of the planet in lower declination, the planet of natally lower declination will be brought into parallel with the one at the natally higher declination. Such aspects are what we term double-ended, i.e., both of the bodies are equally accented. In the first case described the aspect is single-ended and the body that has the parallel made to it by the directed body is the more vital one.

Annual and Monthly Rates

Two ways in which solar arc in longitude can be determined have been described. But with solar declination arc it is very advisable to find its length only by using the ACD. This is because in certain cases of high declination the change in its rate of motion in only one day can make an appreciable difference in the length of the arc. Just as the motion of the Sun in one day is by direction its motion in one year in longitude, so the same is true of its rate of motion in declination. If a given date in an ephemeris is equivalent to the ACD in the current year, then the next day in the ephemeris is equivalent to the ACD of the next year, i.e., by direction, and the difference in declination of the Sun's positions in declination on the two dates—its motion in one day—becomes the annual rate.

When very near the equator and in very low declination, this annual rate may equal 24′, giving a monthly rate of 2′. When near the turn in a very high declination, the annual rate may be only 1′, or even less (in the vicinity of 0° Cancer or 0° Capricorn for the Sun, Ascendant, or Vertex in solar declination arc, Ascendant declination arc, or vertical declination arc). In any case the annual rate divided by 12 gives the monthly rate. One uses the monthly rate to determine how long it will take from the ACD until an exact parallel or contraparallel is reached by direction.

The *average* annual rate of solar declination arc is 0°15′24″, equivalent by declination arc to the measure of Naibod (0°59′08″) the *average* annual rate of the Sun's motion in *longitude*. Whereas the variation from the Naibod value is close to plus or minus 2′—for actual variations in the annual rate of the true solar arc in longitude—the variation of solar declination arc is much greater, i.e., from 15.4′ to a maximum of 24′, and from 15.4′ to a minimum annual rate of nearly 0′. Since the potency of a direction is in inverse proportion to its rate, there is much greater variation in the potency of aspects by declination arc than by arcs in longitude. The smaller the annual rate the more powerful the effects of the parallels and contraparallels, but the fewer of them there will be. The larger (faster) the annual rate the more parallels and contraparallels there will be, but the less potent they will be.

Let us say that we wish to determine the annual rate of either the Ascendant declination arc or the vertical declination arc. They are found in the same way so it will be sufficient to describe only one of the two. It is necessary first to know the annual rate of the Ascendant arc in longitude. Let us assume that it is slow and is 43′30″ per year. And let us also assume that the directed Ascendant is in a part of the zodiac where the Sun has a daily motion in longitude in the ephemeris of just 58′. Then one side of our equation will be 43′30″/58′. Next we find what the daily motion of the Sun is in declination in the same part of the zodiac where the directed Ascendant is located. Let us say that it is 16′. Our answer will be 16′ x (43′30″/58′) or 12′ per year. Or we can state it as a proportion: 43′30″ is to 58′ as X is to 16′. Since 43′30″ is just three-fourths of 58′, 12′ (or X) is just three-fourths or 16′. Then dividing the 12′ annual rate by 12 we find that the monthly rate of the Ascendant declination arc is 1′. Finding these annual rates is important and is no particular problem for those who are used to the process of interpolation.

There are fewer aspects by these three declination arcs than by those in longitude since we use only parallels and contraparallels. But they can be very potent in their effects if the arc is slow. In addition, the work of John Nelson of RCA suggests that exact multiples of three degrees in the planets' distance apart in declination may prove to be significant. This is an entirely new idea that astrologers should test.

Directing the Midheaven by Solar Declination Arc
Customarily we direct the Midheaven by solar arc in longitude, but it may also be directed by solar arc in declination. The position of the Midheaven in longitude is the distance in arc along the plane of the ecliptic to the Midheaven, from where the ecliptic crosses the equator—in particular from the 0°00′ Aries crossing point or equinox. The position of the Midheaven in declination is the distance measured in arc along the plane of the meridian to the equator, which the meridian crosses or cuts at a right angle. If then the Midheaven is moved (directed) both of the values change, i.e., the value of the declination of the Midheaven as well as the value of the longitude. Therefore, conversely, the application of declination arc to the Midheaven as well as the application of solar arc in longitude to it will cause the Midheaven to move, or progress.

To do this one needs first to determine the position in declination of the natal Midheaven (by bringing the Sun in the ephemeris to the same longitude and then finding its declination) and, also, whether its declination is increasing or decreasing. If it is increasing (north or south), one then merely adds the solar declination arc to this natal declination of the Midheaven. This gives its directed declination.

To find its position in longitude, one then finds what longitude the Sun would have if it had the same declination as the directed Midheaven. If the declination of the Midheaven is decreasing at birth, one subtracts the solar declination arc from this value to find the directed declination of the Midheaven, and finds its longitude equivalent as usual. To determine whether the declination of a radical or directed angle (Midheaven, Ascendant, or Vertex) or of a natal or directed Sun is increasing or decreasing, one has only to remember that if they are between 0°00′ Aries and 0°00′ Cancer (if north), or opposite this, i.e., between 0°00′ Libra and 0°00′ Capricorn (if south), they will be increasing; if between 0°00′ Cancer and 0°00′ Libra (if north) or opposite this, i.e., between 0°00′ Capricorn and 0°00′ Aries (if south) they will be decreasing in declination.

The rules given above apply consistently only to the Sun and the three angles.

Once one has found the Midheaven directed by solar declination arc—a second directed Midheaven, the other one being given by solar arc in longitude—one may find the Ascendant that corresponds to this Midheaven and the Vertex that corresponds to this Midheaven (actually, IC), these being the directed positions of these two angles directed by declination. The rates of motion of the three angles by this separate general mode of direction are subject to considerably more variation than is true of the Midheaven directed by solar arc in longitude (and the corresponding directed Ascendant and Vertex). If, for instance, the Midheaven is in a part of the zodiac where the Sun has a daily motion in declination of 24′ (maximum), but the Sun in the chart is actually moving only 1′ a year by solar declination arc, then the motion of the Midheaven will be 1′/24′ times the motion in longitude that the Sun has there. So if the Sun would move 60′ in that part of the zodiac, the Midheaven will move only one twenty-fourth of that rate of 2′30″ a year! An equally great variation in the other direction is possible since the Midheaven may move very rapidly. When the motion is very slow the effects will be very powerful.

To find the annual rates one proceeds as follows: the rate that the Sun moves per year by solar declination arc divided by the rate that the Sun moves per day in declination in the part of the zodiac where the directed Midheaven is located—a fraction—is multiplied times the rate that the Sun moves in longitude in the same part of the zodiac per day (year by direction). The result is the annual rate of the Midheaven in longitude due to its direction by solar declination arc. Having found this longitudinal rate of the Midheaven, that of the Ascendant and Vertex are easily found in the usual manner.

Needless to say, these new arcs can be used most profitably in forecasting future conditions as well as in rectification. I have used them consistently for many years, and they have been tested by our ABC Research Group. We are certain they will produce excellent results if they are tested just as carefully by other astrologers. Solar declination arc may be used even when the time of birth is inaccurate, although if the Sun's natal declination or directed declination is very high, an appreciable error will creep in.

In our opinion, based on a great deal of testing and practical experience, the six modes of direction described in this series—the three arcs in longitude and the related three arcs in declination—together comprise astrology's most valuable timing technique. They are relatively simple methods of directing, especially the arcs in longitude.

In making the transition from these new forms of primary direction to the more usual type, let me briefly describe the basic unity of all these forms of direction. Let us imagine that it is the commencement of spring with the Sun at 0°00′ Aries and that it is true local noon so that the Sun is conjunct the Midheaven. Ten days later the Sun will have moved 9°54′05″ in solar arc of longitude. But ten days later it will also again be true local noon so that the Midheaven will again be conjunct the Sun at 9°54′05″ Aries. The right ascension of the Sun and Midheaven when read along the equator will be 9°05′53″. Therefore the solar arc in right ascension will be that same 9°05′53″. Both arcs will be true descriptions of the motion of both the Sun and Midheaven in that ten days, or ten years by direction. The declination of the Sun and Midheaven will have increased from its 0°00′ value at 0°00′ Aries to 3°55.4′ north. As a result the solar declination arc will be 3°55.4′ for the ten years.

This then is a right spherical triangle of three sides. Each of the three sides is an equally valid arc of direction, i.e., they are not in conflict with each other. Indeed, the hypotenuse side—along the ecliptic in longitude—is the algebraic sum of the other two sides. It is this directional triangle that is the real key to all forms of primary direction. The three arcs above are not dependent for their correctness on the time of birth and so may be used with unknown or very questionable times. This is a great value that they each have. On the other hand they do not take account of the effects of terrestrial latitude or co-latitude, nor of celestial latitude. However, the Ascendant and vertical arcs in both longitude and declination do take account of and are affected by terrestrial latitude and co-latitude. And the declination arcs are also affected by the celestial latitude of the Moon and planets.

Primary directions that involve oblique ascension (of types 1 and 2) for aspects to and by the Ascendant (type 1) and Vertex (type 2) are affected by both terrestrial and celestial latitude values. This is also the case of primary directions under the body's own pole or co-pole. This is one reason why they are so important and should not be neglected even though they require more work. In the final section of the book examples of these various primary directions are given. They are not as difficult as they may seem to be at first. Indeed, once the basic speculum has been made, the rest is much easier.

Chapter 3

Primary Directions

Primary directions were assumed to be based on Earth's rotation. During the first quarter of a day or six hours after birth, the Midheaven shifts some 90°.

Ptolemy suggested that on the basis of one degree equaling one year, this would signify 90 years of life. Then a full day (four times as long) would be equivalent to 4 x 90 years or 360 years of life. If instead of using the symbolic and inaccurate one degree a year, one uses Naibod (360° divided by 365.25 days), or 0°59′08″, then one day by primary direction would equal 365.25 years. In short, one day would equal one year times one year!

A third measure was given by Simmonite such that if at birth the Sun was moving at 58′ a day, then it should have this same constant rate throughout life. The best investigators and technicians have reached the conclusion that we should use the true motion of the Sun instead of Naibod's mean motion or Ptolemy's or Simmonite's keys. Sepharial proposed the true motion (in right ascension varying from 54′ to 66′, and in longitude varying from a little over 57′ to a little over 61′ a day) and it was used by Johndro, DeLuce, Schwickert, Wagner, and others. On the basis of their experience and my own, I assume that only the true motion is valid, which is clear enough when small orbs are used (ranging from 5′ for most aspects to 20° at the most as contrasted with the ridiculous but widely used plus or minus 1°, a range of 120′).

Solar Arc in Right Ascension

On the basis of the investigations of Wagner, Johndro, Schwickert and others we should also reject the older semi-arc approach. Let me first describe the method of primary directions employed by some of their best students. Let me also suggest that you obtain either Hugh Rice's tables of houses, due to its supplementary tables, or the less expensive *Complete Method of Prediction* by Robert DeLuce, due to its equally valuable supplementary tables.

Aspects to and by the Midheaven are based on the true solar arc in right ascension. The right ascension-longitude table in the Appendix helps to find this arc. Or one may simply convert celestial longitude into right ascension. In *Dalton's Tables of Houses* (Placidus) and in Marr's very fine *Campanus Houses*, for each degree of longitude on the Midheaven, the equivalent degrees and minutes of right ascension are given, so it is not hard to convert longitude into right ascension sine (celestial) latitude. The Sun and Midheaven have no latitude.

But in order to use this equatorial arc, as DeLuce terms it, one must convert all one's natal longitudes into their equivalent right ascensions. Since the Moon and planets usually do have celestial latitudes to make such conversions, one needs the table in Rice or DeLuce (given the celestial longitude and latitude to find the equivalent right ascension—see the Appendix for a description of double interpolation and the right ascension longitude table. Or one does this by trigonometry.

Johndro's birth locality (BL) charts are directed, as to their Midheavens, by both this right ascensional (true) solar arc, which is added to the R.A.M.C. of the chart, and also by the longitude of the Midheaven to which is added the (true) solar arc in longitude. However, you are warned not to use the baseline (#1) given in his book *The Earth in the Heavens* (1929) since in articles in Margaret Morrell's *Modern Astrology* of 1950-1951, he shifted that baseline to a static one (#2) such that the Greenwich meridian has a right ascension of 0°00′ and *not* the 29°10′ (of January 1, 1930) of his first baseline! Gustav Schwickert in his fine *Rectification* has shown the use of this true solar arc in right ascension in the successful rectification of natal Midheavens using 5′ orbs. But like Johndro he also recognized the equal validity of solar arc in longitude. A body that forms a conjunction to the Midheaven via right ascension has its body right on the meridian plane which is usually not true of a conjunction in longitude (unless it has no celestial latitude).

Aspects of the Ascendant

For a body to be bodily on the plane of the horizon it must make the conjunction in oblique ascension (or descension). The difference between the right ascension and oblique ascension is termed its ascensional difference. Given the terrestrial geocentric latitude of birth and the natal declination of any body, one may find its ascensional difference from an ascensional difference table, using double interpolation, such as one finds in Rice or DeLuce, or one may find it by trigonometry. In some cases the ascensional difference is added to and in other cases subtracted from the right ascension to obtain the natal oblique ascension or oblique descension. The laws governing this are given on page 33 of DeLuce. (The laws of when to add or subtract the ascensional differences are also given on page 48. The formula for finding the ascensional difference by trigonometry is also given there.)

Bodies on the rising side of the chart have oblique ascension and on the setting side have oblique descension. The oblique ascension of the Ascendant is always 90° plus the R.A.M.C. Once more the right ascension solar arc is used for aspects to or by the Ascendant, but from their oblique ascensional and oblique descensional positions and not their right ascensional positions, and with the Ascendant's oblique ascension also. Let us say that you wish to bring Uranus to the square of the Ascendant and that this is on the rising side of the chart. Then the arc from the radical oblique ascension of Uranus to the oblique ascension of its square to the Ascendant is the length of the equatorial solar arc.

Aspects of the Vertex

In taking aspects to or by the Vertex one employs oblique ascension and oblique descension of type 2, which are found by adding ascensional difference 2 to, or subtracting it from, the natal right ascension. One may use an ascensional difference table, and enters it with the radical declination of the body at the co-latitude of the place.

From the prior chapter you may recall that the co-latitude = 90° - latitude (geocentric). If you had found the oblique ascension 1 by adding its ascensional difference 1 to its right ascension, then its oblique ascension 2 = right ascension - ascensional difference 2, i.e., you now subtract as the sign is always opposite. Or had you subtracted the ascensional difference 1, you would now add the ascensional difference 2. Oblique ascension 1 and oblique descension 1 positions, while read on the equator, correspond to positions on the prime vertical. Oblique ascension 2 and oblique descension 2 positions, while read on the equator, correspond to places on the horizon (in azimuth). The oblique ascension of the Vertex is the R.A.M.C. - 90°. So much then for aspects to and by the three angles. What then of bodies not conjunction or opposition any one of these three angles? Aspects between such bodies must be taken under the pole of the significator. Due to errors in the determination of said "own pole," errors in timing have contributed to the neglect of these primary directions.

Directions *in Mundo* Under Own Pole and Co-pole

The polar elevation of the North Pole above the horizon in the northern hemisphere is always equal to the (geocentric) latitude of the place. This is the polar elevation at the horizon and of the Ascendant. But from there the polar elevation decreases to a 0° value at the upper and lower meridians (Midheaven and IC). Thus a body that lies between the horizon and upper or lower meridian (as most do) has its own pole which is greater than 0° but less than the terrestrial latitude of the place.

This own pole value has been miscalculated by Zadkiel, Sepharial (who copied him), DeLuce (who used a different method but got the same result), and Keuhr. The correct determination of the own pole I have taken from Cyril Fagan as published in the *AFA Journal of Research* (1947).

One passes a great circle through the north and south points of the horizon that also passes through the body of the planet. The place where this great circle cuts the equator is its modified oblique ascension 1 (or oblique descension 1) under its own pole. I term this its vertical ascension. The place where this same great circle cuts the ecliptic, I term its vertical longitude. The place where it cuts the prime vertical, which it always must do perpendicularly, I term its zenith distance, i.e., as measured along the prime vertical from the zenith. This zenith distance position is also its position in Fagan's Mundoscope. Then the sine of P1 = sine of Phi x dine of ZD (zenith distance). Phi is the terrestrial latitude and P1, the polar elevation. When zenith distance is 90° (the arc from the zenith to the horizon and in the Campanus house system, and therefore, the distance from the Midheaven to the Ascendant along the prime vertical) then P1 is equal to the latitude of the place; when ZD = 0°, then P1 = 0° also. One uses the value of P1 instead of the latitude of the place in order to find the modified ascensional difference 1 in the ascensional difference table (using the usual declination also). Then the modified Q1 (ascensional difference 1) plus or minus the right ascension gives the vertical ascension (or descension).

Latitude of place (geocentric) = Phi. Difference in right ascension between the R.A.M.C. and the right ascension of the body (from the upper or lower meridian, whichever is the closer one) is the meridian distance (MD).

TAN A = COS PHI X TAN MD
TAN B = TAN PHI X COS MD

If north declination and MD taken from Midheaven or if south declination and MD taken from IC, then B - dec1ination = C.

If south declination and MD taken from Midheaven or if north declination and MD taken from IC then B + declination = C.

TAN D = SINE PHI X SINE MD X TAN C

Then ZD = A + D unless declination is north and exceeds B, and MD is taken from Midheaven, or unless declination is south and exceeds B, and MD is taken from IC. In these two cases AZD = A - D. One should use the logarithms of the trigonometric functions.

The value of zenith distance may also be found to the nearest tenth of a degree (nearest 0°) by the use of Hydrographic Office publication #214 *Tables of Computed Altitude and Azimuth*, via triple interpolation (described in the Appendix), but no trigonometry is necessary. #214 gives the azimuth to the nearest 0.1° and the altitude to the nearest $0.1'$ and is intended mainly for and is widely used by navigators. One needs only: Phi, declination, and meridian distance. The values in the table are given in nine volumes for every degree of Phi from 0° to 90° (North or South Pole). The values of the declination are given for every $0^{\circ}30'$. The values of the Midheaven are given for every 0° of right ascension.

One may find the positions on the prime vertical (zenith distance) instead of along the horizon in azimuth by using the co-latitude in place of the latitude and by taking the meridian distance from the opposite meridian in using #214. The tables will not give values for bodies within less than five degrees of either the horizon or the prime vertical. Since they will not give bodies below the horizon (and therefore only on one side of the prime vertical), one reverses the meridian from which the meridian distance is measured (upper to lower or vice-versa) and changes the sign of the declination, i.e., puts the body on the other side of the zodiac. One then finishes by adding or subtracting 180° to obtain the correct result.

Very similar methods enable one to find the own co-pole, or P2. Phi′ is the co-latitude (geocentric). Therefore, sine P2 = Sine Phi′ X sine Z, where Z is the azimuth or distance measured along the horizon from the north point of it as 0°. If one uses #214, one uses the table in the usual way, i.e., entering it at Phi (the latitude and not the co-latitude) and taking the meridian distance from the closest upper or lower meridian. If the body is below the horizon, one changes the sign of the declination and shifts the meridian distance to the opposite meridian and then adds 180° to the final result.

If one uses the Fagan trigonometric formulae, one substitutes Phi' for Phi (i.e., uses the co-latitude in place of the latitude) and shifts the meridian distance to the other meridian. Having found P2, one then uses it to enter the ascensional difference table to find Q2, the "modified" ascensional difference 2. Then the radical right ascension plus or minus this Q2 gives the horizontal ascension (or descension) which is used in place of oblique ascension 2 and oblique descension 2 for aspects between planets that are not conjunction or opposition any of the local angles. The use of horizontal ascension with the usual solar equatorial arc has been developed by me.

Directions Under Own Pole and Co-pole *in Zodiaco*
The best oblique ascension-descension table is that of Kuehr. It assumes the obliquity of the ecliptic to be just 23°27′00″ and gives the longitudes for every one degree of oblique ascension and oblique descension and for every degree of ter-

restrial latitude except in the upper 40's and lower 50's of latitude where values are given for every 0°10′ of latitude. Positions are given to the nearest 1″ and not merely the nearest 1′. Once one has found the vertical ascensions (or descensions) and horizontal ascensions (or descensions) that correspond on the equator to places on first the prime vertical and then on the horizon, one may use this table to find the vertical longitudes and horizontal longitudes for these bodies under their own poles and co-poles. The procedure is as follows:

Given a vertical ascension, one finds the equivalent longitude at the latitude of birth (or other place). This is the vertical longitude. If one has instead a vertical descension, one must add 180° to it (or subtract 180° from it), find the vertical longitude, and add six signs of 180° to obtain the correct result. Given a natal horizontal ascension, one finds the longitude by adding or subtracting 180°, and by using the radical co-latitude. Then to the result one adds six signs, or 180°, to obtain the correct horizontal longitude. But if one has a natal horizontal descension, one need not add or subtract 180°, although one must use the co-latitude to find the horizontal latitude from the table.

Say one wishes to bring Saturn to a square of the Moon using horizontal longitude values and using solar, Ascendant or vertical arc in longitude. One finds the horizontal longitude of Saturn under its own co-pole and then the horizontal longitude of a point square to the Moon, also under the co-pole of Saturn (as it is the significator). Then the arc between these values will be one of the longitudinal arcs named above.

Furthermore one may take the horizontal longitude of the place of the aspect either sine latitude or under the (celestial) latitude of the significator (Saturn). If we use the latitude of Saturn, then in finding the modified ascensional difference 2 (used for finding its horizontal ascension) we employ the co-pole 2 and the declination it would have with this latitude and longitude of the place of aspect. If the place of aspect is taken in the ecliptic (sine latitude), then the declination will be of that longitude but without any latitude. Such an essential table for declinations will be found in DeLuce's book.

Johndro noted the vital importance of these vertical longitude and horizontal longitude points as did Wagner, but they otherwise for the most part have been generally overlooked. In my tests of them they appear to be quite important. It is work to find them, but once one has found the pole and co-pole of the planets the rest is not very difficult. It is to be hoped that other astrologers will also test them. The use of horizontal longitude and vertical longitude values does much to remove the need for the wide orbs that most astrologers find necessary.

Just as the vertical longitude and horizontal longitude positions on the ecliptic correspond to the vertical ascension (and descension) and horizontal ascension (and descension) positions on the equator, so do the oblique longitude 1 and oblique longitude 2 positions on the ecliptic correspond to oblique ascension 1 (or descension 1) and oblique ascension 2 (or descension 2) positions on the equator. Oblique longitude (or type 1) is found through the Kuehr table from oblique ascension l (or descension 1) just as from the same table we found vertical longitude from vertical ascension (or descension). Oblique longitude (of type 2) is found at the co-latitude from the table and oblique ascension 2 (or descension 2) just as we found horizontal longitude in the same table from horizontal ascension (or descension). One uses ecliptic arcs—solar, Ascendant, and vertical arc in longitude—with oblique longitude l positions for aspects to and by the Ascendant's longitude. Similarly one uses oblique longitude 2 positions for aspects to and by the Vertex. Oblique longitude l, oblique longitude 2, vertical longitude and horizontal longitude all take account of celestial latitude effects, which is not true of the conventional zodiacal longitude positions. When there is no celestial latitude, these four positions disappear into zodiacal longitude.

Just as we may use oblique longitude 1and oblique longitude 2 positions for aspects by and to the Ascendant and Vertex, so we may use right longitude positions for aspects along the ecliptic to and by the Midheaven. Having found the right ascension, we look up equivalent longitude in a right ascension to longitude table (Rice, DeLuce, or Kuehr's other volume). Right longitude will only differ from zodiacal longitude, however, if the body has celestial latitude. Below are corresponding positions on the equator and ecliptic:

Equatorial Positions	*Ecliptic Positions*
Right Ascension (RA)	Right Longitude (RL)
Zodiacal Ascension (ZA)	Zodiacal Longitude (ZL)
Oblique Ascension l (OA1)	Oblique Longitude 1 (OL1)
Oblique Ascension 2 (OA2)	Oblique Longitude (OL2)
Vertical Ascension (VA)	Vertical Longitude (VL)
Horizontal Ascension (HA)	Horizontal Longitude (HL)

A great circle through the North and South Poles of the ecliptic that passes through a planet cuts the ecliptic at its zodiacal longitude. Where it cuts the equator is the zodiacal ascension of that body. Two bodies having the same zodiacal ascension points are in conjunction in longitude (zodiacal longitude) just as two bodies that have the same right longitude positions are in conjunction in right ascension. The trigonometric formula for finding the value of zodiacal ascension when zodiacal longitude is given is as follows: log co-tangent of ZA = log co-tangent of ZL + log cosine of Omega. Here Omega is the obliquity of the ecliptic approximated as 23°27′. The log cosine of 23°27′ is 9.9622562. The value of ZL is to be measured from either 0° Aries or 0° Libra, i.e., either preceding or following either 0° Aries or 0° Libra (so that ZL is less than 90°). Then ZA's value in RA will result, as less than 90°, measured backward or forward from the same 0° Aries or 0° Libra. These ZA and RL values used by Johndro appear in his very important books: *The Earth in the Heavens* and *The Stars, How and Where They Influence*. Fagan and Johndro must be accounted as two of the most important technical astrologers of the 20th century.

Johndro's Primary Directions

Johndro's most advanced and later conclusions on locality shifting and on primary directions have never been published. Furthermore they are closely related.

Johndro found the right ascensions of all natal bodies, then their ascensional differences 1 and ascensional differences 2, and from this their natal oblique ascensions 1, oblique descensions 1, oblique ascensions 2, and oblique descensions 2. He then used the usual true solar arc in right ascension (converse as well as direct) to shift the radical bodies to their directed positions. As far as I know he did not determine their vertical ascensions and horizontal ascensions. The directed positions were then converted to longitudinal values by means of an oblique ascension table (such as the Kuehr table). In other words, the directed oblique ascension 1 and oblique descension 1 positions were converted into directed oblique longitude 1 values at the latitude (geocentric) of birth, or other latitude. And the directed oblique ascension 2 and oblique descension 2 positions were converted at the radical co-latitude to directed oblique longitude 2 positions. Then the aspects to the natal bodies were taken from these directed oblique longitude 1 and oblique longitude 2 positions.

The annual rates of said directed oblique longitude 1 and oblique longitude 2 bodies were obtained by finding the positions one year later, then taking the difference between oblique longitude 1 for this year and oblique longitude 1 for next year. The annual rates of the various bodies by this method are not identical, but vary somewhat, i.e., Mars may be moving more rapidly than Saturn.

Let us say that one wants to know when Jupiter will be trine Neptune by his oblique longitude 2 mode of direction (along the horizon but read on the ecliptic). One finds the value of the oblique descension 2 (assuming that Neptune is on the setting side of the chart) which corresponds to the longitude of a point on the ecliptic trine natal Neptune at the co-latitude of birth. This then will be the directed value of the oblique descension 2, which directed Jupiter must reach.

Let us assume that natal Jupiter is rising in the first house. Then we find its right ascension, its ascensional difference 2, which we add to or subtract from its right ascension to obtain its oblique ascension 2. Then the difference between this oblique ascension 2 of Jupiter and the oblique descension 2 of Neptune's trine position will give the required solar equatorial arc.

Both Fagan and Johndro insisted on using a ratio of the sidereal day to the year (1:366.25) instead of the usual solar day to the year (1:365.25). If one uses their ratio, 60 days after birth is equivalent to 60 years plus 60 days of life instead of the more usual 60 years and no days. I prefer and use the conventional ratio on both theoretical and empirical grounds.

Other Effects of Celestial Latitude

One may say that orbs are a sign of ignorance. Johndro first demonstrated the significance of this in his two books (already mentioned). One may read an aspect between the Sun and Jupiter in the field plane of Mars. Given the zodiacal longitude and celestial latitude of a body, we can from a table (DeLuce, Rice) find the equivalent right ascension. Let us say that we use the longitude of Jupiter but the latitude of Mars and then find the equivalent right ascension; we are finding the right ascension of Jupiter in the field plane of Mars.

In his two books Johndro gives a number of examples of such cases and shows that they reduce orbs to value of from 0′ to 2′! Already the various right, vertical, horizontal, oblique 1 and 2 longitudes have been described as accounting for the effects of celestial latitude, all of which disappear into zodiacal longitude when there is no celestial latitude (sine latitude). This approach of Johndro's is still another vital and valid way of taking account of celestial latitude and in so doing making timing in astrology a precision matter. Actually a given aspect may operate in all the field planes of the other eight bodies, although not in exactly the same manner nor of course at the same time. The period from the effect of the aspect in the first field plane to that of the last one may well be the basis of any real orb!

After having mastered several modes of progression and direction, especially secondary progression and the new directions in longitude and the solar arc in right ascension, you are encouraged to study my *The Technique of Rectification* (included in *The Best of Charles Jayne*) since there is no more vitally important technique for any astrologer.

Two diagrams show what has been described to this point. The diagram on page 50 shows the celestial sphere looking from the east toward the west. An attempt has been made to show the vertical ascension, vertical longitude, right ascension and right longitude positions of a body above the horizon and having celestial latitude above the ecliptic. Note that the east point is the oblique ascension 1of the Ascendant.

The diagram on page 51 shows the celestial sphere looking from the west toward the east. The place shown has the same terrestrial latitude (40°) in both diagrams and both of them have the same Midheaven (0°00′ Aries). On this second diagram we show the horizontal descension and horizontal longitude points of a body with north celestial latitude and south declination. Also shown are the two points on the ecliptic parallel to that body.

Mundane Arcs and Aspects

Previously the inaccurate semi-arc system was used with the equally faulty Placidus house system. No technically informed astrologer uses the Placidus system. Campanus houses were used by Fagan, Johndro, Carter, and Rudhyar. I use them. Arcs along parallels of declination are, as Wagner has indicated, along minor rather than great circles, which invalidates their accuracy. Only Harwood had a way of correcting for this error, i.e., through the use of his polar angles.

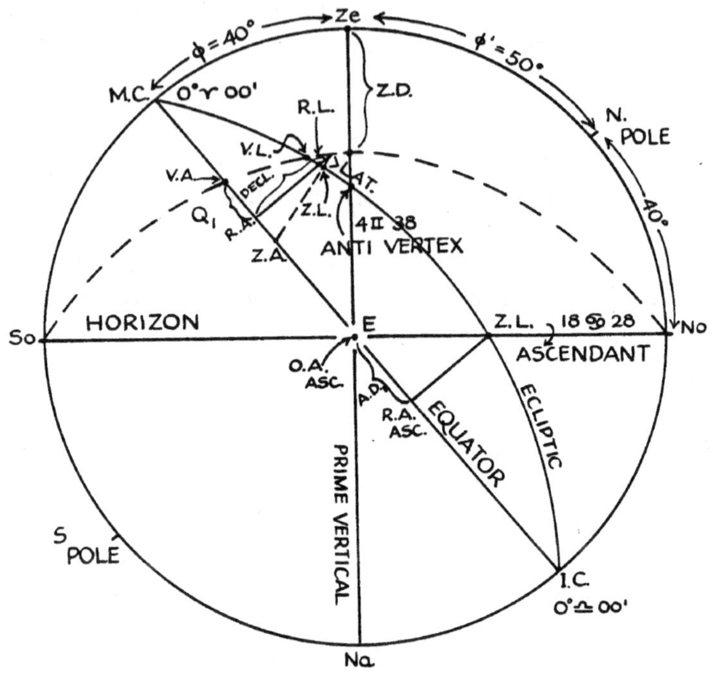

In the Mundoscope of Fagan, projected on the prime vertical in place of the ecliptic, the interplanetary aspects are mundane rather than zodiacal (on the ecliptic). In the Zenithscope of Zieggruen, the positions of bodies are projected on the horizon (in azimuth), and again the aspects are mundane instead of zodiacal.

You may have Saturn trine your Sun zodiacally, but square mundanely. But you can move to another locality where the mundane square disappears (the radical ecliptic trine will always remain unchanged), so that in the new locality it is much easier to give tangible expression (house position and mundane aspect) to your inner saturnine capabilities (sign position and zodiacal aspect).

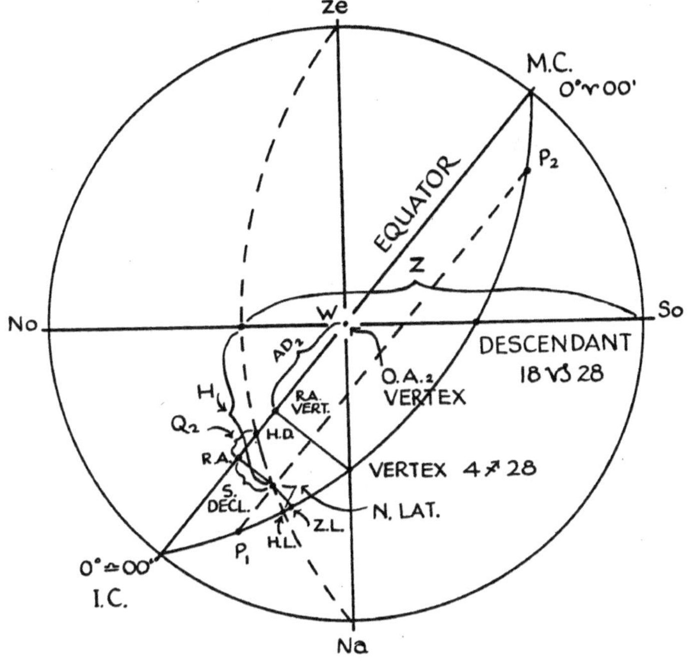

There should, therefore, be mundane arcs of direction, i.e., prime vertical arcs and azimuth arcs which could be used to energize these natal mundane aspects. These would be new mundane arcs and not the old discredited ones. At this time they are still under investigation and could be a fruitful line of inquiry. As of now we have quite enough modes of direction without having to add more!

Chapter 4

Symbolic Directions

From my forty years experience I find that that which is astrologically significant always has an astronomical basis. For this reason I tend to look askance at all symbolic methods (degrees, directions, names, etc.) in astrology. Many years ago Eleanore Hesseltine, one of the great unknown astrologers of the twentieth century, made a very thorough test of the several methods of symbolic direction proposed by Charles Carter and Frankland. She had detailed diaries and extraordinary skill in validation, rectification, and prediction techniques. She concluded that for the most part they were invalid or minor.

There was one measure, however, that she found to be valid and that was one-quarter degree a year. But it can be shown that this has an astronomical basis. A sidereal year is 365.256 days in length; a tropical year (peculiar to the tropical zodiac) is 365.242 days in length. There is about a quarter of a day more than 365 days in a year. And in that quarter day Earth rotates close to 15′. Should one use .256 or .242? At the end of about 71.5 years there is a difference of one degree and since about one-quarter degree is equivalent to one year of life, the difference amounts to four years!

Should we take one-quarter of the mean motion or the true motion? Certainly we should use the true motion. Let us say that your solar arc (longitude) in 40 years is 39°020′. Divide 39°20′ by that 40 days to find your mean daily rate over the 40 years. It will be 59.0′. Then to find your quarter degree rate, multiply that 59.0′ by either .256 or .242. I have not done extensive work on this arc but thus far the sidereal .256 value appears to be better. One can just hear the hosanna from the sidereal zodiac camp! This quarter turn of Earth is also important in shifting the Midheaven of a solar return (sidereal or tropical) from year to year. It is also the key to Wynn's (Sidney K. Bennet) Key Cycle mode of timing. Thus it has not gone unrecognized as to its importance by astrologers.

To find your one-quarter degree arc, multiply your solar arc in longitude by either .256 or .242, whichever you prefer, and the result will be your quarter degree arc of direction, which I really must refuse to describe as "symbolic"! It should be just as valid as a converse arc or a direct one. Indeed, I suggest that for converse directional arcs we use the term conversions. The quarter degree arc should not be applied to either the Ascendant or the Vertex, whose motion is always in some ratio to the motion of the Midheaven. We also can multiply the .256 (or .242) ratio times the solar arc in right ascension in order to obtain the quarter degree arc in right ascension as well as in longitude.

Summary

Basically we have described three related arcs plus this fourth special quarter degree one. The first of the three arcs is solar and has three forms: in longitude, in right ascension, and in declination. The second one is the Ascendant arc in longitude and declination. The actual motion of the Ascendant in oblique ascension 1 is the same length as the right ascension solar arc so that the Ascendant arc has only two forms. The same then is so of the vertical arc, i.e., two forms, or in longitude and declination. There may well be a mundane prime vertical arc and mundane azimuth arc. The three related arcs basically derive from the three local static planes and their angles and thus from the three dimensions of space. The three dynamic and cosmic planes—ecliptic, equator, and lunar plane—give us, through the three periods (solar or annual, diurnal, and lunar or 27.32 day period, and their ratios to each other), the three modes of progression.

Chapter 5

Examples

Let us take the case of a man born October 9, 1911 at 75W08, 40N05 (geographic) at a given time of 10:43 pm EST. The time was rectified to 10:39:30 pm and has been proven accurate for twenty-two years. First we correct the geographic latitude to geocentric. From it one takes 0°11′ and subtracts it from the 40N05 latitude to obtain the geocentric of 39N54. The rectified time gives the following values:

Body	*Long*	*Lat*	*RA*	*Decl*	*RLS*
Sun	15 ♎ 43	0:00	194:28	-6:11	15 ♎ 43
Moon	11 ♉ 28	+0:57	38:43	+16:06	11 ♉ 09
Jupiter	16 ♏ 40	+0:50	224:26	-16:02	16 ♏ 54
Uranus	25 ♑ 25	-0:33	297:30	-21:37	25 ♑ 32
Pluto	28 ♊ 59	-6:27	88:50	+17:00	28 ♊ 56

This is the first part of what is termed a speculum. Only half of the bodies have been included since just a sample is being given. The longitudes, latitudes, and declinations are taken directly out of any good ephemeris. The right ascensions are determined by single interpolation from a table that converts longitude and latitude (celestial) to right ascension. Single interpolation is explained in the Appendix so will not be given here. The right longitude (RL) is taken from a right ascension longitude table. One does not take account of celestial latitude in finding the longitude (right) that corresponds to the right ascension. Note that the Sun's RL and (conventional) ZL are the same. This is because the Sun has no celestial latitude.

Next we list the three Angles:

Angle	*Longitude*	*RA*	*Declination*
Midheaven	27 ♓ 09	357:23	-1:08
Ascendant	16 ♋ 14	107:37	+22:28 terr. lat. (Phi) + 39:50
Vertex	2 ♐ 35	240:30	-20:42 terr. co-lt. (Phi) +50:06

As the angles have no latitude (sine lat.), the latitude column is not given, nor is the RL column as their ZLs and RLs are the same. We also need to know the ACD for the secondary progressions. So to 10:39:30 pm EST we add the five hours necessary to find GMT, which gives 3:39:30 am GMT of the next day (October 10, 1911). Now from then until noon GMT is 8:20:30. Each 2 hours equals one month by progression so that 8 divided by 2 (8/2) = 4 months. And each 4 minutes of time equals 1 day of life so that 20′30″/4 = 5 days. By progression (secondary) then the 8 hours and 20′30″ are equivalent to 4 months 5 days which are added to the date of birth to obtain the ACD since Greenwich noon is later. Had we gone back to the prior Greenwich noon (or October 9) this would have been equivalent to an ACD of February 14, 1911 instead of February 14, 1912. It won't matter which one we use!

In the 1911 ephemeris, Jupiter is at 27°09′ Scorpio on November 27 (noon G.M.T.) in exact trine to the 27°09′ Pisces Midheaven. This is just 48 days after Greenwich noon of October 10, which is equivalent to February 14, 1912. By progression this must be 48 years later, so it falls on February 14, 1960. Within a few days of that date the native received national publicity in *LIFE* magazine and four months earlier (October 1959) had sponsored an international astrological conference in the United States. In June 1970 this same aspect occurred by RA (secondary progressed RA of Jupiter was trine the R.A.M.C. of 357:23 from a RA of 237:23). Five months before that in January 1970 the native was a co-sponsor of another international astrological conference in the United States, and in March 1970 received some national publicity in *McCall's*.

To time this, one finds from the longitude and latitude of progressed Jupiter as of the ACD of February 14, 1970 its RA (longitude at this time—58 days after October 10 or on December 7 was 29°21′ Scorpio and the latitude was +0:45), which was 237:18.2. One does the same thing for the next day in the ephemeris (December 8) when the longitude was 29°35′ Scorpio and the latitude was still +0:45. This then gives a progressed RA for February 14, 1971 of 14.5′ more than it had been at February 14, 1970 and thus its annual rate. To reach 237:23 (actually 237:23.25) it must move from 237:18.2 to 237:23.2 or 5′. Then this divided by its annual rate of 14.5′ times 12 months gives the interval after February 14, 1970 that the aspect became exact or 4 months—June

1970. It is suggested that astrologers make use of those RA aspects to the Midheaven by secondary progression.

Let us bring the Midheaven to a trine of Uranus by solar arc in longitude. The Midheaven must move from 27°09′ Pisces to 25°25′ Taurus, which is 58°16′. Let us add this to the ephemeris Sun for October 9, 1911 at 15°04′19" Libra which will give us 13°20′19″ Sagittarius. On December 6 (58 days later) the Sun will be at 13°15′36″ Sagittarius. This is equivalent to 58 years after the birthday plus an arc of 4′43″ to move from 13°15′36″ to 13°20′19″ Sagittarius. The monthly rate is nearly 5.1′ so that it will take about 27 days to move the 4.7′ and the aspect will then occur in early November 1969. One week later the native flew to Atlanta, Georgia and back in one day with a business associate on a successful business trip paid for by the company.

The Midheaven must move another 7′ to reach a trine to the RL of Uranus at 25°32′ Capricorn, which is one month and 12 days later or a little past the middle of December 1969. About a week earlier he and the same associate made the same trip (overnight) to the same place by air and with regard to the same business matter. In fact they were seeing that something was "right" on this right longitude aspect! In late July 1971 on the Midheaven trine Uranus by solar arc in right ascension the native will again fly to Atlanta, Georgia on a teaching trip (two-day) at the expense of others. It is striking that the trips are to the same place. The rulership of Uranus over the Aquarian eighth house shows that the trips are at the expense of others. The aspects to the Midheaven show that they are business trips.

In Chapter 2 is a table of the Sun's right ascension for each day. The RA of the Sun in the horoscope is 194°28′. Uranus' natal RA is 297°30′15″, so the Midheaven must move 60°07′ to trine Uranus from 57°30′15″. Adding the 60°07′ to the natal Sun's RA we obtain 254°35′. The most accurate way to find the time this occurs is to convert it into longitude. In the tables of RA, 254°35′ lies between 253°43′ (equivalent to 15o Sagittarius) and 253°47′ (equivalent to 16° Sagittarius). This gives us 42′/64′ = x/60′ (254°35′ - 253°43′ = 42′; 254°47′ - 253°43′ = 64'; 16°00′ Sagittarius - 15°00′ Sagittarius = 60′). So in longitude the primary Sun is at 15°39′4″ Sagittarius. This is 59.5 days (years) after the ACD so is exact mid August 1970, about 2.5′ after the event. The above method is more accurate than DeLuce's apparent RA table. Since the daily (annual) rate of the solar RA arc is more variable than the solar longitude arc, it is essential to use the ACD and the radical Sun rather than the ephemeris Sun in finding the RA solar arc.

Next let us take the case of a female born June 20, 1923 with a natal Midheaven of 29°26′ Aries, Ascendant at 10°36′ Leo, Phi at 40°27′ (geocentric), the Sun being at a longitude 28°10′ Gemini, the Moon at 11°04′ Virgo and singleton Uranus at 17°33′ Pisces in the natal eighth house and ruling the sign Aquarius in the seventh house. Below is the data essential for our further examples.

Midheaven: 29 Aries 26 (long); 0:00 (lat); 27:22 (RA)
Ascendant: 10 Leo 36 (long); 0:00 (lat); 117:22 (OA1); 10 Leo 36 (OL1)
Sun: 28 Gemini 10 (long); 0:00 (lat); 88:00 (RA); +23:26 (decl); -21:42 (AD); 66:19 (OA1) 28 Gemini 10 (OL1)
Uranus: 17 Pisces 33 (long); -0:47 (lat); 348:52 (RA); -5:38 (decl); -4:40 (AD); 344:02 (OA1); 17 Pisces 18 (OL1)

There are two additional places at which bodies operate on the ecliptic. Consider the declination of Uranus at -5:38. Bring the Sun to that declination, using interpolation (single) and find what longitude it will have. This will be the point on the ecliptic parallel to the planet.

In this case, it is at 15°43′30″ Pisces or near its zodiacal longitude. This is a parallel longitude position; there are always two of them as they are equidistant from 0° Cancer and 0° Capricorn on the ecliptic and thus have the same declination. 15°4′30″ Pisces is 0o00′ Capricorn plus 75°43′30″. The other parallel longitude (parallel longitude 2; parallel longitude 1 is always the one closest to the zodiacal longitude) is then 0°00′ Capricorn - 75°43′30″ = 14°16′30″ Libra.

Using Ascendant arc we bring the Ascendant to a contraparallel of Uranus or to a point in opposition to parallel longitude 1 at 15°43′30″ Virgo. The Midheaven for such an Ascendant is at 13°26′ Gemini. Then the timing is given by the distance of solar arc in longitude of the natal Midheaven (29°26′ Aries) to this directed Midheaven, i.e., 13°26′ Gemini - 29°26′ Aries = 44°00′. The ACD of this chart is June 6, 1923 so that 44°00′ plus the natal Sun = 12°10′ Leo.

On August 5, 1923, it is at 12°01′43″ Leo (46 days or 46 years by progression after the June 6, 1923 ACD). The additional arc of 8.3′ is to be measured via the daily (annual) rate of the Sun, which is 57.5, i.e., (8.3′/57.5′) x 12 months = 1.73 months or 1 month plus 22 days. Thus the aspect occurred at the very end of July 1969, half a week before her husband was notified that he and 200 other people had been laid off from their jobs. Note Uranus in the eighth house of the husband's income.

Let us say that we wish to bring the Ascendant to an opposition of Uranus by equatorial arc. The oblique ascension 1 of the Ascendant (R.A.M.C. + 90°) is brought to an opposition of the oblique descension 1 of Uranus (the asterisk means that in the oblique ascension 1 column the value for Uranus is actually an oblique descension 1 value). Subtract 180° (the opposition) from Uranus' oblique descension 1 (344:02), which gives 164:02. The oblique ascension 1 of the Ascendant is 117:22. The right ascension solar arc is 164:02 - 117:22 = +46°40′. We add this to the natal right ascension of the Sun: 46°40′ + 88:00 = 134°50′. In DeLuce's right ascension longitude tables (or any other such table) a right ascension of 134°50′ is equivalent to 12°21′ Leo. In an ephemeris, August 5 has the Sun at 12°01.7 Leo and on August 6 it is at 12°59.2′, a difference of 57.5′. Then 12°21.0′ - 12°01.7′ = 19.3′, so we have 19.3′/57.5′ = x/12 months, making the aspect in early October 1969. Thus for about two months the husband's income was erratic and the effect was nerve-wracking.

The first aspect had been Ascendant contraparallel Uranus, but it also was Uranus by Ascendant declination arc contraparallel the Ascendant! In short it was a double-ended aspect and is cited as an example from life of an Ascendant declination arc. In order to see an example of an oblique longitude aspect at work, let us take this same case and bring the Ascendant by Ascendant arc (longitude) to an opposition of the oblique longitude 1 of Uranus at 17°18′30″ Pisces. The directed Midheaven for an Ascendant at 17°18′30″ Virgo is 15°12′30″ Gemini. Then the solar arc is 15°12′30″ Gemini less the 29°26′ Aries of the natal Midheaven = +45°46′30″. Added to the natal Sun this gives 13°56′30″ Leo. At the ACD of June 5, 1971 the Sun is at 13°56′42″ Leo. Thus the aspect occurred in early June 1971 at another time of marked financial stress for the native due to her husband's income, but also at a time of becoming separated from certain people who had been friends. In October 1971 the Ascendant opposed Uranus' zodiacal longitude. During the weeks just before this a financial deal over land with a friend unexpectedly fell through due to unforeseen medical expenses in connection with his mother.

Johndro Primaries

Let us now take Jupiter square natal Sun in the chart of the man. We shall consider that at the place of aspect, the aspect is taken in the ecliptic (sine latitude). This being the case both the vertical longitude and horizontal longitude of the place of aspect will be exactly square the Sun's zodiacal longitude, i.e., natal Sun at 15°43′ Libra and aspect at 15°43′ Capricorn. Jupiter being on the setting side of the horoscope, we are dealing with descensions and not with ascensions. This affects the use of the Kuehr and #214 tables. From the first part of the speculum we know that the right ascension of Jupiter is 224:26 and that its declination is -16:02 (minus means south and + means north declination).

Therefore we look up the ascensional difference l in the ascensional difference table for the terrestrial latitude (Phi) of 39°54′ and find the ascensional difference l to be -13:54. (Minus since it is on the setting side and has southern declination. Also minus if on the rising side with north declination; otherwise the ascensional difference is added. These laws reverse in the southern hemisphere). Thus the oblique descension l is 210:32. To do the aspect by the Johndro method we next must find the oblique descension l of the place of aspect (descension since still on the setting side of the chart). Using the table we find the oblique descension l for a longitude of 15°43′ Capricorn at a Phi of +39°54′. But we must do a trick with the table when dealing with either oblique descension l or oblique ascension 2 values (need not do this with oblique ascension l or oblique descension 2 values). We find the oblique ascension l for 15°43′ Cancer (i.e., subtract six signs from the 15°43′ Capricorn value), which proves to be 86:37. Then add or subtract 180o to this oblique ascension l to find the oblique descension 1, which becomes 266:37. Then 266:37 - 210:32 = +56°04′, which is the right ascension solar arc. Added to the right ascension of the natal Sun it gives 250°32′. Via interpolation in a right ascension longitude table this is equivalent to a longitude of 12°03′ Sagittarius. This turns out to be a little more than two months before his 56th ACD, or early December 1967.

Let us say that we wish to do the oblique descension 2 aspect. Then using the same declination but a Phi of 50°06′ (co-latitude) we find that the ascensional difference 2 is +20:06; it is + as the ascensional difference 1 was minus (they always have the opposite sign). The oblique descension 2 of Jupiter is 244:32. Now we must find what oblique descension 2 15°43′ of Capricorn has, i.e., at a Phi of 50°06′. When using the table with an oblique descension 2 we need not add six signs. We obtain an oblique descension of 316°47′. The needed arc is then 316°47′ - 244°32′ = +72°15′ (direct arcs are +; converse arcs are -). What we have just described is the Johndro method of primary arcs.

Primary Directions Under Own Pole or Co-pole in Mundo and *in Zodiaco*

The polar elevation of a body may be found in any of three ways: from the trigonometric formula already given (from Fagan), from H.O. #214, and from H.O. #211 (Ageton's method). The Fagan method and #211 are both trigonometric and must be used if one wants exact values, or a body is within fewer than five degrees of the horizon of prime vertical. H.O. #211 is quite inexpensive and contains its own secant and co-secant tables. There is a very clear explanation of how to use the tables to obtain the azimuth and altitude in this small book. By using the co-latitude and taking the MD (meridian distance in right ascension, which in this book is termed the LHA or local hour angle) from the opposite meridian one may also find the zenith distance and amplitude (the amplitude is the perpendicular distance in arc that a body is on either side of the prime vertical). Therefore I shall not try to explain how to use H.O. #211 here. By following the steps of the Fagan method given in the primary directions section, one can easily use this other trigonometric method. The use of H.O. #214 is given in an Appendix, and an example is also given there of how to turn the zenith distance into the pole, and the azimuth into pole and co-pole. I shall, therefore, assume that we have found the pole and co-pole of Jupiter, which in the case of the man are 26°57′ north (PI) and 47°07′ north (P2 or co-pole). Our task is now to list the items in the speculum:

Jupiter: 26:57 (P1); -8:24 (Q1); 216:02 (VD); 17 Scorpio 28(VL); 47:07 (P2); +18:01 (Q2); 242:27 (HD); 16°06′30″ Scorpio (HL)

We find the modified ADl (termed Ql) with the usual declination of -16:02, but with PI instead of Phi in an AD table. We then find the modified AD2 (called Q2) with the usual declination, but with P2 instead of Phi (47°07′ instead of 50°06′). Since we subtracted the unmodified ADl to find OD1, we subtract Ql from the RA of 224°26′ to find VD. To bring Jupiter to a square of the Sun by equatorial arc under its own pole we must now find the VD value at the place of the aspect, which we shall do sine latitude.

In order to do this we must find the Q1 at the place of aspect. Therefore we need PI (pole of Jupiter) and the declination of 15°43′ Capricorn. From an ephemeris we find that this point on the ecliptic (as though the Sun were there) has a South declination of -22:32. From this the Q1 of the place of aspect is -12:11. From the RA table, the RA of 15°43′ Capricorn (sine latitude) is 277:46.25. Jupiter VD 286:47.25 - 12:11 = 277:46.25. 277:46.25 - 216:02 = + 58°44′15″ RA solar arc. Natal Sun 194:28′ + 58°44.25′ gives directed Sun in table 253:12.25, which is equivalent to late May 1970. At that time the native was overly optimistic about his financial outlook regarding a job that started at the beginning of May 1970 and as to the profitability of a publication (Jupiter) that was to come out at the same time, and also as to the division of some land (Sun rules Leo in his second house and is natally in his fourth house).

VL and HL aspects are easier to do since one need not find the VL and HL of the place of aspect as one must do with VA - VD and HA - HD positions unless the aspect is taken with latitude (of the significator—in this case, Jupiter). Thus from a VL of 17°28′30″ Scorpio to 15°43′ Capricorn is + 58°14′30″.

By simple solar arc in longitude this occurred in late October 1969. At that time a very expensive apartment was rented in a nearby large city. Then the lease was cancelled but at a loss.

This fits well with a square of Jupiter that involves the second and fourth houses. From the HL to 15°43′ Capricorn is an arc (elliptical) of 59°36′30″, which hit at the end of February 1971. At that time a 95-acre property was found that was a good buy, but that was too expensive, although a real bargain in realty values. Again this fits, and whereas the city apartment was large and fancy, here the land was more than ample, all typical of Jupiter. From the ZL of 16°40′ Scorpio the solar arc is 59°03′ which hit in early August 1970.

At that time there was an extremely well-paid business project that should have gone through but didn't. At the same time an effort was made to buy a house and to sell the current one, but in spite of looking very promising at the time, it did not work out, partly because the house sought would have been too expensive. I should remark that these VL and HL positions appear to work out quite significantly and accurately as to time. They are prime vertical and horizontal positions that are read on the ecliptic. If the place of aspect is taken sine latitude, they are easy to do once the basic VL and HL values have been found.

If one wishes to replace the aspect with the latitude of the significator, one must find the declination of the place of aspect at the latitude of the significator, and then with this declination and the pole (or co-pole) of the significator, one finds the VA or VD (or HA or HD) of the place of aspect. Then one finds the equivalent VL and HL places.

Appendix

Single Interpolation

Let us say that our calculations show that the sidereal time of a birth is 18:40:19 and that we wish to find the Midheaven.

We find in a tables of houses for a Midheaven of 9o00' Capricorn that the ST is 18:39:11 and that for the next Midheaven at 10°00′ Capricorn the ST is 18:43:31. The difference between the two STs is (18:43:31 - 18:39:11 =) 0:04:20. Turning that into seconds of sidereal time we have 260 for the denominator. Then from the 18:39:11 to our given ST is 18:40:19 - 18:39:11 = 0:01:08, which in seconds of ST is 68. The 260 is always the denominator and the 68 is the numerator. 68 seconds of ST is to 260 seconds of ST as X is to 60' of arc of longitude or 68/260 = x/60. This becomes x = (60 x 68)/260. We simply divide both numerator and denominator by 20; i.e., (3 x 68)/13 = (204/13) = 15.7′. This then is added to the earlier Midheaven at 9°00′ Capricorn to give a Midheaven of 9°15.7′ Capricorn, which rounds off to 9o16' Capricorn. The above is single interpolation which is constantly used in astrology.

Double Interpolation

Now we will find the Ascendant for this Midheaven. The

geocentric latitude of the chart is 52°35′ north. This requires double interpolation, for you are not only finding an Ascendant for a value between two Midheavens but also between two latitudes (i.e., 52° and 53° north).

First do a single interpolation between the two Midheavens —one at 52° north and the other at 53° north. After that we interpolate between the two latitudes in order to obtain the final value.

In other words we simply do the two interpolations one after another so that it really is not complicated.

	Midheaven	9° ♑	9°15.7	10 ♑
Lat.				
53°		24°21′ ♈	25°00.25	26°51′ ♈
52°35′			24°33.9 Answer	
52°		23°19′ ♈	23°56.9	25°44′ ♈

For accuracy we shall retain the Midheaven as 9°15.7′ Capricorn. At 52° north our equation is 15.7/60 = x/2°25′. The 2°25′ (145’) is the distance between the two Ascendants (for the two Midheavens) at 52o latitude.

Solving, we find that we must add 37.9′ to the 23°19′ Aries Ascendant and this gives 23°56.9′ Aries. At 53° the equation is 15.7/60 = Y/2°30′. For at 53° the two Ascendants (for the two Midheavens) are farther apart than they are at 52°. Solving (reduce degrees and minutes to minutes first, or 2°30′ to 150′) we have to add 39.25′ to the earlier Ascendant at 53°, i.e., 24°21′ Aries + 0°39.25′ = 25°00.25′ Aries.

Now we can do the second interpolation. The interval between the two latitudes is, as always, just 60′ (the denominator). We wish to go from 52° to 52°35' or 35′, which becomes the numerator of one side of the equation. The denominator of the other side of the equation will be the difference between the two interpolated Ascendants or 25°00.25 - 23°56.9′ = 0°63.35′. Then the equation is 35/60 = z/63.35. This gives +37.0′ to be added to the Ascendant at 52°, i.e., 23°56.9′ Aries + 0°37.0′ = 24°33.9 Aries, which is to the nearest whole minute, and thus to our answer, 24°34′ Aries.

We shall soon see that triple interpolation is no more difficult, simply requiring one additional step. In other words, you always do just one interpolation at a time and in this way can do more if need be.

Triple Interpolation

Usually this will only be necessary with such a table as H.O. #214. Given the terrestrial latitude and co-latitude, the declination, the meridian distance (MD which = either RAMC - RA of body, or, if nearer, RAIC - RA of body). First we shall determine the azimuth (Z) for which #214 was intended. If the body is above the horizon the table is used in a straightforward manner.

On the left-hand pages of #214 are the values for Z (and H, the altitude) for "Declinations Same Name as Latitude" and on the right-hand pages are values of Z and H for "Declinations Contrary Name to Latitude." "Name" means arithmetic sign (+ being north and - being south) and by latitude the terrestrial kind is meant. The values of Z are in the left columns in bold type and the values of H are in the right columns in light type.

Let us find the Z for the Moon in the chart of the female native. Its RA is 162:26, which places it closer to the IC than the

Midheaven, and its MD will be the RAIC or 207:22 less the RA of the Moon and thus is 44:56. Its declination is +7:07, and it is at 11°04′ Virgo and thus below the Horizon (the Ascendant being at 10°36′ Leo). Normally we could not find Z from this table if the body were below the horizon, but we may do so by putting the Moon on the opposite side of the zodiac (in Pisces) and thus above the horizon. This makes its MD refer to the MC instead of the IC and it changes the sign of its declination to -7°07′. As she was born at 40°27′ north geocentric terrestrial latitude, the declination and latitude have opposite signs and we must look on the right-hand page (Contrary Name).

First we shall interpolate between MDs of 44° and 45° for a MD of 44°56′ at both a declination of -7°00′ and one of -7°30′, all at a Phi of 40°00′. Then having found the Zs at each of the two declinations we make our second interpolation between the two declinations for the correct -7°07′ value. Below is the first or 40°00′ "box" of the first two interpolations.

MDs	*Declinations*		
	-7°00′	-7°07′	-7°30′
44°00′	128.7°		129.0°
44°56′	127.86°	127.93°	128.16°
45°00′	127.8°		128.1°

Singly underlined values have to do with single interpolation, and doubly underlined ones with the second interpolation. The 127.93° is used with the final value in the 41°00′ "box" for the third interpolation.

Next we repeat the whole process, only now at a latitude

41°00′, which is on a different (but still right hand) page of #214. Below is the 41°00′ "box" of values.

MDs	*Declinations*		
	-7°00′	-7°07′	-7°30'
44°00′	129.1°	____	129.4°
44°56′	128.26°	128.33°	128.56°
45°00′	128.2°	____	128.5°

The third interpolation is between 40o and 41o for the correct value of 40°27′ and between 127.93° (Z at 40°) and 128.33° (Z at 41°). The final result is 128.11° rounded off to 128.1°.

The second decimal place is retained throughout until the very end when the result is rounded off to the nearest first decimal value. As our "shift" put the Moon (in Pisces) closer to the Midheaven than to the IC the 128°06′ must be subtracted from 180° to put it back on the correct side of the celestial sphere.

This gives 51°54′ as the value of Z and its distance from the north point of the horizon which is always the 0o value in the northern hemisphere. This gives the Moon's position on the horizon in the Zenithscope. This value of Z used with the value of the co-latitude (here 49°33′ north) is then used to find the co-pole (P2). That is to say, that the log sine of P2 = log sine Phi (co-latitude) + log sine Z.

But let us say that we wish to find the zenith distance (ZD) measured along the prime vertical (i.e., at right angles to the horizon) from the zenith. We started out with measuring MD, in this case from the IC (before our "shift" for calculation purposes) and used #214 with the latitude (Phi). We may use this same table to find the ZD if we do to it what we do to a tables of houses in finding the Vertex. Turn the chart upside down—MD now measured from the Midheaven instead of the IC. Use the co-latitude (Phi) instead of the latitude. Having done this we go through the same triple interpolation process as we did to find the azimuth. We are turning Z into ZD and H (altitude) into amp (amplitude, the perpendicular arc of distance the body is on either side of the prime vertical plane north or south of it). Let us say that you follow this procedure only to discover that there are no values for the ZD in the table. You then do just what you did in finding the Z when you were on the wrong side of the horizon, i.e., you shift the MD to the opposite meridian and you change the sign of the declination.

Then when you get the result be sure to subtract it from 180°. In this case the ZD comes out to be 123°30′ from the zenith or 56°30′ from the closer Nadir (both measured along the prime vertical). Then the log sine of Pl = log sine Phi + log sine ZD. Pl for this Moon is +32°45′ and its P2 = +36°46′30″ or the co-polar elevation of the Moon. Draw small, rough diagrams in order to avoid making mistakes. The ZD value is used also to give the position of a body along the prime vertical in the Mundoscope and in Campanus houses.

Use of Right Ascension Tables

This explains how to find the right ascension, given the celestial longitude and the celestial latitude.

The table on the following pages is given for both north and south celestial latitudes to 6°00′. By doing double interpolation one may find the right ascension. Venus does, however, at times reach a latitude of about 8°45′ and Pluto can reach about 18° north and south. In such cases the following trigonometric formula may be used. It also enables one to find the declination for a given celestial longitude and celestial latitude.

Let L stand for longitude, lt for latitude and Ob for the obliquity of the ecliptic (exactly 23°27′ in 1917). RA is right ascension and dec. the declination.

Log sine L (from 0° Aries or 0° Libra) (or log cosine from 0° Cancer or 0° Capricorn) + log tangent Ob - log tangent angle A.

If Land lt are both positive (L measured in the order of the signs) or if both are negative, then 90° - lt - angle B.

If L and lt are of different name (opposite signs) then 90° + lt - B.

Angle B - angle A - angle C

Log cosine A + log cosine C + log cosine Ob - log sine of dec. (In the first term one uses the arithmetical complement, which is the log subtracted from 10.0000000.)

If angle B is greater then 90°, then the declination will have the opposite sign ("contrary name") to the sign of the celestial latitude.

Log cosine dec. (arithmetical complement) + log cosine L (Aries or Libra) (or log sine L from Cancer or Capricorn) + log cosine lt log cosine RA (from Aries or Libra) (or log sine RA from Cancer or Capricorn).

To Find the Ascensional Difference

Log tangent of the pole + log tangent of the dec. - log sine AD. For the oblique ascension for aspects to the Ascendant, the pole is the terrestrial geocentric latitude (Phi) of the place.

The OA (or OD) is the AD added to or taken from the RA. The laws for doing this are as follows:

RA - AD = OA if dec. is north
RA + AD = OA if dec. is south
RA + AD = OD if dec. is north
RA - AD = OD if dec. is south

This is true for the northern hemisphere, but the signs are reversed in the southern hemisphere.

From the Oblique Ascension or Descension to Find the Ecliptic Longitude

The superb tables of Kuehr (one for RA to longitude, the other volume for OA or OD to longitude) are now out of print.

From Alfred J. Pearce's *The Textbook of Astrology*, here is the formula for converting OA and OD into ecliptic longitude:

Log cosine OA (from Aries or Libra; log sine OA if from Cancer or Capricorn) + log cotangent pole = log cotangent angle X.

If OA is less than 90° or over 270° angle X + Ob = angle Y
If OA is over 90° but less than 270° angle X - Ob - angle Y

Then log cosine angle Y (arithmetical complement must be used) + log cosine angle X + log tangent OA (if from Aries or Libra, otherwise log cotangent if from Cancer or Capricorn) = log tangent longitude from Aries or Libra (or log cotangent from Cancer or Capricorn).

If the angle Y is greater than 90°, take the log sine of the excess over 90° instead of the log cosine, using of course the arithmetical complement in the first term. If angle Y does exceed 90°, the longitude will fall the reverse way from the point from which the OA was taken.

In all of the above what is said applies also to the OD.

If the place of birth is in the southern hemisphere, the rules are reversed in the third part, i.e., for southern hemisphere. If OA is greater than 90o but less than 270°, add Ob to angle X. If OA is less than 90° or larger than 270°, subtract Ob from the angle X to find the angle Y.

Every competent astrologer should have some knowledge of logarithms and of basic trigonometry. With tables of both logarithms and of the trigonometric functions many problems are easily solved.

N. Lats.	*0*	*1*	*2*	*3*	*4*	*5*	*6*
♈	00 00	359 37	359 13	358 49	358 25	358 01	357 37
01	00 55	00 32	00 08	359 44	359 20	358 56	358 32
02	01 50	01 27	01 03	00 39	00 15	359 51	359 27
03	02 45	02 22	01 58	01 34	01 10	00 46	00 22
04	03 40	03 17	02 58	02 29	02 05	01 41	01 17
05	04 35	04 12	03 48	03 24	03 00	02 36	02 12
06	05 30	05 07	04 43	04 19	03 55	03 31	03 07
07	06 25	06 02	05 33	05 14	04 50	04 26	04 02
08	07 21	06 57	06 33	06 09	05 45	05 21	04 57
09	08 16	07 52	07 28	07 04	06 40	06 16	05 52
10	09 11	08 47	08 23	07 59	07 35	07 11	06 47
11	10 06	09 42	09 18	08 55	08 31	08 07	07 43
12	11 02	10 38	10 14	09 51	09 27	09 03	08 39
13	11 57	11 33	11 09	10 46	10 22	09 58	09 34
14	12 53	12 29	12 05	11 42	11 18	10 54	10 30
15	13 48	13 25	13 01	12 38	12 14	11 50	11 26
16	14 44	14 20	13 57	13 34	13 10	12 46	12 22
17	15 40	15 16	14 53	14 30	14 06	13 42	13 18
18	16 35	16 12	14 49	15 26	15 02	14 39	14 15
19	17 31	17 08	16 45	16 22	15 58	15 35	15 11
20	18 27	18 04	17 41	17 18	16 54	16 31	16 07
21	19 23	19 00	18 37	18 14	17 51	17 28	17 04
22	20 20	19 56	19 33	19 11	18 48	18 25	18 01
23	21 16	20 53	20 30	20 08	19 45	19 22	18 58
24	22 12	21 50	21 27	21 05	20 42	20 19	19 55
25	23 09	22 47	22 24	22 02	21 39	21 16	20 52
26	24 06	23 44	23 21	22 59	22 36	22 13	21 50
27	25 02	24 41	24 19	23 57	23 34	23 11	22 48
28	25 59	25 38	25 16	24 54	24 31	24 09	23 46
29	26 57	26 35	26 13	25 51	25 29	25 07	24 44

N. Lats.	*0*	*1*	*2*	*3*	*4*	*5*	*6*
♉	27 54	27 33	27 11	26 49	26 27	26 05	25 42
01	28 51	28 30	28 08	27 47	27 25	27 03	26 40
02	29 49	29 27	29 06	28 45	28 23	28 01	27 38
03	30 46	30 25	30 04	29 43	29 21	28 59	28 37
04	31 44	31 23	31 02	30 41	30 19	29 58	29 36
05	32 42	32 21	32 00	31 39	31 18	30 57	30 35
06	33 40	33 20	32 59	32 38	32 17	31 56	31 34
07	34 38	34 18	33 58	33 37	33 16	32 55	32 33
08	35 37	35 17	34 57	34 36	34 15	33 54	33 33
09	36 36	36 16	35 56	35 36	35 15	34 54	34 33
10	37 34	37 15	36 55	36 35	36 15	35 54	35 33
11	38 33	38 14	37 54	37 35	37 15	36 54	36 33
12	39 33	39 14	38 54	38 35	38 15	37 55	37 34
13	40 32	40 13	39 54	39 35	39 15	38 56	38 35
14	41 31	41 13	40 54	40 35	40 16	39 57	39 36
15	42 31	42 18	41 54	41 36	41 17	40 58	40 38
16	43 31	43 18	42 54	42 36	42 18	41 59	41 39
17	44 31	44 13	43 55	43 37	43 19	43 00	42 40
18	45 31	45 14	44 56	44 38	44 20	44 01	48 42
19	46 32	46 14	45 57	45 39	45 21	45 03	44 44
20	47 32	47 15	46 58	46 40	46 23	46 05	45 49
21	48 33	48 16	47 59	47 42	47 25	47 07	46 49
22	49 34	49 17	49 00	48 44	48 27	48 09	47 52
23	50 35	50 18	50 02	49 46	49 29	49 12	48 55
24	51 36	51 20	51 04	50 48	50 32	50 15	49 58
25	52 38	52 22	52 06	51 51	51 35	51 18	51 02
26	53 40	53 24	53 09	52 54	52 38	52 22	52 06
27	54 42	54 27	54 12	53 57	58 42	53 26	53 10
28	55 44	55 29	55 15	55 00	54 45	54 30	54 14
29	56 46	56 32	56 18	56 03	55 49	55 34	55 18

N. Lats.	*0*	*1*	*2*	*3*	*4*	*5*	*6*
♊	57 48	57 35	57 21	57 07	56 53	56 38	56 23
01	58 51	58 38	58 24	58 10	57 57	57 42	57 28
02	59 53	59 41	59 27	59 14	59 01	58 47	58 33
03	60 56	60 44	60 31	60 18	60 05	59 52	59 38
04	61 59	61 47	61 35	61 22	61 10	60 57	60 44
05	63 03	62 51	62 39	62 27	62 15	62 02	61 50
06	64 06	63 55	63 43	63 32	63 20	63 08	62 56
07	65 09	64 59	64 47	64 37	64 25	64 13	64 02
08	66 13	66 03	65 52	65 42	65 30	65 19	65 08
09	67 17	67 07	66 57	66 47	66 36	66 25	66 14
10	68 21	68 11	68 02	67 52	67 42	67 31	67 21
11	69 25	69 16	69 07	68 57	68 48	68 38	68 28
12	70 29	70 21	70 12	70 08	69 54	69 45	69 35
13	71 34	71 26	71 17	71 09	71 00	70 51	70 42
14	72 38	72 31	72 22	72 15	72 06	71 58	71 49
15	73 43	73 36	73 28	73 21	73 13	73 05	72 57
16	74 47	74 41	74 33	74 27	74 19	74 12	74 04
17	75 52	75 46	75 39	75 33	75 26	75 19	75 12
18	76 57	76 51	76 45	76 39	76 33	76 27	76 20
19	78 02	77 56	77 51	77 45	77 40	77 34	77 28
20	79 07	79 02	78 57	78 52	78 47	78 41	78 36
21	80 12	80 08	80 03	79 59	79 54	79 49	79 44
22	81 17	81 13	81 09	81 05	81 01	80 56	80 52
23	82 22	82 18	82 15	82 11	82 08	82 04	82 00
24	83 28	83 24	83 21	83 18	83 15	83 11	83 09
25	84 33	84 30	84 27	84 25	84 22	84 20	84 17
26	85 38	85 36	85 33	85 32	85 29	85 28	85 25
27	86 44	86 42	86 40	86 39	86 37	86 36	86 34
28	87 49	87 48	87 46	87 46	87 44	87 44	87 42
29	88 55	88 54	88 53	88 53	88 52	88 52	88 51

N. Lats.	*0*	*1*	*2*	*3*	*4*	*5*	*6*
♋	90 00	90 00	90 00	90 00	90 00	90 00	90 00
01	91 05	91 06	91 07	91 07	91 07	91 08	91 09
02	92 11	92 12	92 14	92 14	92 15	92 16	92 18
03	93 16	93 18	93 20	93 21	93 23	93 24	93 26
04	94 22	94 24	94 27	94 28	94 30	94 32	94 35
05	95 27	95 30	95 33	95 35	95 38	95 40	95 43
06	96 32	96 36	96 39	96 42	96 45	96 48	96 51
07	97 38	97 42	97 45	97 49	97 52	97 56	98 00
08	98 43	98 47	98 51	98 55	99 00	99 04	99 08
09	99 48	99 52	99 57	100 01	100 07	100 12	100 16
10	100 53	100 58	101 03	101 08	101 14	101 19	101 24
11	101 58	102 04	102 09	102 15	102 21	102 26	102 32
12	103 03	103 09	103 15	103 21	103 27	103 33	103 40
13	104 08	104 14	104 21	104 27	104 34	104 41	104 48
14	105 13	105 19	105 27	105 33	105 41	105 48	105 56
15	106 17	106 24	106 33	106 39	106 47	106 55	107 03
16	107 22	107 29	107 38	107 45	107 53	108 02	108 11
17	108 26	108 34	108 43	108 53	108 59	109 09	109 18
18	109 31	109 39	109 48	109 57	110 05	110 15	110 25
19	110 35	110 44	110 53	111 03	111 12	111 22	111 32
20	111 39	111 49	111 58	112 08	112 18	112 29	112 39
21	112 43	112 54	113 03	113 13	113 24	113 35	113 46
22	113 47	113 57	114 08	114 18	114 30	114 41	114 52
23	114 51	115 01	115 18	115 23	115 35	115 47	115 58
24	115 54	116 05	116 17	116 28	116 41	116 52	117 04
25	116 57	117 09	117 21	117 33	117 46	117 58	118 10
26	118 01	118 13	118 25	118 38	118 51	119 03	119 16
27	119 04	119 16	119 29	119 42	119 55	120 08	120 22
28	120 07	120 19	120 33	120 46	120 59	121 13	121 27
29	121 09	121 22	121 36	121 50	122 03	122 18	122 32

N. Lats.	*0*	*1*	*2*	*3*	*4*	*5*	*6*
♌	122 12	122 25	122 39	122 53	123 07	123 22	123 37
01	123 14	123 28	123 42	123 57	124 11	124 26	124 42
02	124 16	124 31	124 45	125 00	125 15	125 30	125 46
03	125 18	125 33	125 48	126 03	126 18	126 34	126 50
04	126 20	126 36	126 51	127 06	127 22	127 38	127 54
05	127 22	127 38	127 54	128 09	128 25	128 42	128 58
06	128 24	128 40	128 56	129 12	129 28	129 45	130 02
07	129 25	129 42	129 58	130 14	130 31	130 48	131 05
08	130 26	130 43	131 00	131 16	131 33	131 51	132 08
09	131 27	131 44	132 01	132 18	132 35	132 53	133 11
10	132 28	132 45	133 02	133 20	133 37	133 55	134 14
11	133 28	133 46	134 03	134 21	134 39	134 57	135 16
12	134 29	134 47	135 04	135 22	135 40	135 59	136 18
13	135 29	135 47	136 05	136 23	136 41	137 00	137 20
14	136 29	136 47	137 06	137 24	137 42	138 01	138 21
15	137 29	137 47	138 06	138 24	138 43	139 02	139 22
16	138 29	138 47	139 06	139 25	139 44	140 03	140 24
17	139 28	139 47	140 06	140 25	140 45	141 04	141 25
18	140 28	140 46	141 06	141 25	141 45	142 05	142 26
19	141 27	141 46	142 06	142 25	142 45	143 06	143 27
20	142 26	142 45	143 05	143 25	143 45	144 06	144 27
21	143 25	143 44	144 04	144 24	144 45	145 06	145 27
22	144 23	144 43	145 03	145 24	145 45	146 06	146 27
23	145 22	145 42	146 02	146 23	146 44	147 05	147 27
24	146 20	146 40	147 01	147 22	147 43	148 04	148 26
25	147 18	147 39	148 00	148 21	148 42	149 03	149 25
26	148 16	148 37	148 58	149 19	149 41	150 02	150 24
27	149 14	149 35	149 56	150 17	150 39	151 01	151 23
28	150 11	150 33	150 54	151 15	151 37	151 59	152 22
29	151 09	151 30	151 52	152 13	152 35	152 57	153 20

N. Lats.	0	1	2	3	4	5	6
♍	152 06	152 27	152 49	153 11	153 33	153 55	154 18
01	153 04	153 25	153 47	154 09	154 31	154 53	155 16
02	154 01	154 22	154 44	155 06	155 29	155 51	156 14
03	154 58	155 19	155 41	156 03	156 26	156 49	157 12
04	155 54	156 16	156 39	157 01	157 24	157 47	158 10
05	156 51	157 14	157 36	157 58	158 21	158 44	159 08
06	157 48	158 10	158 33	158 55	159 18	159 41	160 05
07	158 44	159 07	159 30	159 51	160 15	160 38	161 02
08	159 40	160 04	160 27	160 49	161 12	161 35	161 59
09	160 37	161 00	161 23	161 46	162 09	162 32	162 56
10	161 33	161 56	162 19	162 42	163 06	163 29	163 53
11	162 29	162 52	163 15	163 38	164 02	164 25	164 49
12	163 25	163 43	164 11	164 34	164 58	165 21	165 45
13	164 20	164 44	165 07	165 30	165 54	166 18	166 42
14	165 16	165 40	166 03	166 26	166 50	167 14	167 38
15	166 12	166 36	166 59	167 22	167 46	168 10	168 34
16	167 07	167 31	167 55	168 18	168 42	169 06	169 30
17	168 03	168 27	168 51	169 14	169 38	170 02	170 26
18	168 58	169 23	169 46	170 09	170 33	170 57	171 21
19	169 54	170 18	170 42	171 05	171 29	171 53	172 17
20	170 49	171 13	171 37	172 01	172 25	172 49	173 13
21	171 44	172 08	172 32	172 56	173 20	173 44	174 08
22	172 39	173 03	173 27	173 51	174 15	174 39	175 03
23	173 35	173 58	174 22	174 46	175 10	175 34	175 58
24	174 30	174 53	175 17	175 41	176 05	176 29	176 53
25	175 25	175 48	176 12	176 36	177 00	177 24	177 48
26	176 20	176 43	177 07	177 31	177 55	178 19	178 43
27	177 15	177 38	178 02	178 26	178 50	179 14	179 38
28	178 10	178 33	178 57	179 21	179 45	180 09	180 33
29	179 05	179 28	179 52	180 16	180 40	181 04	181 28

N. Lats.	*0*	*1*	*2*	*3*	*4*	*5*	*6*
♎	180 00	180 23	180 47	181 11	181 35	181 59	182 23
01	180 55	181 18	181 42	182 06	182 30	182 55	183 18
02	181 50	182 13	182 37	183 01	183 25	183 49	184 13
03	182 45	183 08	183 32	183 56	184 20	184 44	185 08
04	183 40	184 03	184 27	184 51	185 15	185 39	186 03
05	184 35	184 58	185 22	185 46	186 10	186 34	186 58
06	185 30	185 54	186 18	186 42	187 06	187 30	187 53
07	186 25	186 49	187 13	187 37	188 01	188 25	188 48
08	187 21	187 44	188 08	188 32	188 56	189 20	189 43
09	188 16	188 39	189 03	189 27	189 51	190 15	190 38
10	189 11	189 34	189 58	190 22	190 46	191 10	191 33
11	190 06	190 29	190 53	191 17	191 41	192 05	192 28
12	191 02	191 25	191 48	192 13	192 36	193 00	193 23
13	191 57	192 20	192 43	193 08	193 31	193 55	194 18
14	192 53	193 16	193 39	194 03	194 26	194 50	195 13
15	193 48	194 12	194 35	194 58	195 21	195 45	196 08
16	194 44	195 07	195 30	195 53	196 16	196 40	197 03
17	195 40	196 02	196 25	196 48	197 11	197 35	197 58
18	196 35	196 58	197 21	197 44	198 07	198 30	198 53
19	197 31	197 54	198 17	198 40	199 02	199 25	199 48
20	198 27	198 50	199 13	199 36	199 58	200 21	200 43
21	199 23	199 46	200 09	200 32	200 54	201 16	201 39
22	200 20	200 42	201 05	201 28	201 50	202 12	202 34
23	201 16	201 38	202 01	202 24	202 46	203 08	203 30
24	202 12	202 35	202 57	203 20	203 42	204 04	204 26
25	203 09	203 31	203 53	204 16	204 38	205 00	205 21
26	204 06	204 28	204 50	205 12	205 34	205 56	206 17
27	205 02	205 25	205 47	206 09	206 30	206 52	207 13
28	205 59	206 22	206 43	207 05	207 26	207 48	208 09
29	206 57	207 19	207 40	208 01	208 22	208 44	209 05

N. Lats.	*0*	*1*	*2*	*3*	*4*	*5*	*6*
♏	207 54	208 16	208 37	208 58	209 19	209 40	210 01
01	208 51	209 13	209 34	209 55	210 16	210 37	210 57
02	209 49	210 10	210 31	210 52	211 13	211 34	211 54
03	210 46	211 07	211 28	211 49	212 10	212 31	212 51
04	211 44	212 05	212 25	212 46	213 07	213 27	213 47
05	212 42	213 03	213 23	213 43	214 04	214 24	214 44
06	213 40	214 01	214 21	214 41	215 01	215 21	215 41
07	214 38	214 59	215 19	215 39	215 58	216 18	216 38
08	215 37	215 57	216 17	216 37	216 56	217 15	217 35
09	216 36	216 56	217 15	217 35	217 54	218 13	218 32
10	217 34	217 54	218 13	218 33	218 52	219 11	219 29
11	218 33	218 53	219 12	219 31	219 50	220 09	220 27
12	219 33	219 52	220 11	220 30	220 48	221 07	221 25
13	220 32	220 51	221 10	221 28	221 46	222 05	222 23
14	221 31	221 50	222 09	222 27	222 45	223 03	223 21
15	222 31	222 50	223 08	223 26	223 44	224 02	224 19
16	223 31	223 49	224 07	224 25	224 43	225 00	225 17
17	224 31	224 49	225 06	225 24	225 42	225 59	226 15
18	225 31	225 49	226 06	226 23	226 41	226 58	227 14
19	226 32	226 49	227 06	227 23	227 40	227 57	228 13
20	227 32	227 49	228 06	228 23	228 39	228 56	229 12
21	228 33	228 50	229 06	229 23	229 39	229 55	230 11
22	229 34	229 50	230 06	230 23	230 38	230 54	231 10
23	230 35	230 51	231 06	231 23	231 38	231 53	232 09
24	231 36	231 52	232 07	232 23	232 38	232 53	233 08
25	232 38	232 53	233 08	233 24	233 38	233 53	234 08
26	233 40	233 55	234 09	234 24	234 38	234 53	235 07
27	234 41	234 57	235 11	235 25	235 39	235 53	236 07
28	235 43	235 58	236 12	236 26	236 40	236 54	237 07
29	236 46	237 00	237 14	237 27	237 41	237 54	238 07

N. Lats.	*0*	*1*	*2*	*3*	*4*	*5*	*6*
♐	237 48	238 02	238 15	238 29	238 42	238 55	239 07
01	238 51	239 04	239 17	239 30	239 43	239 55	240 07
02	239 53	240 06	240 19	240 31	240 44	240 56	241 08
03	240 56	241 09	241 21	241 33	241 45	241 57	242 09
04	241 59	242 11	242 23	242 35	242 46	242 58	243 09
05	243 03	243 14	243 25	243 37	243 48	243 59	244 10
06	244 06	244 17	244 28	244 39	244 50	245 01	245 11
07	245 09	245 20	245 31	245 41	245 52	246 02	246 12
08	246 13	246 23	246 34	246 44	246 54	247 04	247 13
09	247 17	247 27	247 37	247 47	247 56	248 06	248 15
10	248 21	248 30	248 40	248 49	248 58	249 07	249 16
11	249 25	249 34	249 43	249 52	250 00	250 09	250 17
12	250 29	250 38	250 46	250 55	251 03	251 11	251 19
13	251 34	251 42	251 49	251 58	252 05	252 13	252 21
14	252 38	252 46	252 53	253 01	253 08	253 15	253 23
15	253 43	253 50	253 57	254 04	254 11	254 18	254 25
16	254 47	254 54	255 01	255 07	255 14	255 20	255 27
17	255 52	255 58	256 05	256 11	256 17	256 22	256 29
18	256 57	257 03	257 09	257 15	257 20	257 25	257 31
19	258 02	258 07	258 13	258 18	258 23	258 28	258 33
20	259 07	259 12	259 17	259 21	259 26	259 31	259 35
21	260 12	260 17	260 21	260 25	260 29	260 34	260 38
22	261 17	261 21	261 25	261 28	261 32	261 36	261 40
23	262 22	262 25	262 29	262 32	262 35	262 39	262 42
24	263 28	263 30	263 33	263 36	263 39	263 42	263 45
25	264 33	264 35	264 37	264 40	264 42	264 45	264 47
26	265 38	265 40	265 41	265 44	265 45	265 48	265 49
27	266 44	266 45	266 46	266 48	266 49	266 51	266 52
28	267 49	267 50	267 50	267 52	267 52	267 54	267 54
29	268 55	268 55	268 55	268 56	268 56	268 57	268 57

N. Lats.	*0*	*1*	*2*	*3*	*4*	*5*	*6*
♑	270 00	270 00	270 00	270 00	270 00	270 00	270 00
01	271 05	271 05	271 05	271 04	271 04	271 03	271 03
02	272 11	272 10	272 10	272 08	272 08	272 06	272 06
03	273 16	273 15	273 14	273 12	273 11	273 09	273 08
04	274 22	274 20	274 19	274 16	274 15	274 12	274 11
05	275 27	275 25	275 23	275 20	275 18	275 15	275 13
06	276 32	276 30	276 27	276 24	276 21	276 18	276 15
07	277 38	277 35	277 31	277 28	277 25	277 21	277 18
08	278 43	278 39	278 35	278 32	278 28	278 24	278 20
09	279 48	279 43	279 39	279 35	279 31	279 26	279 22
10	280 53	280 48	280 43	280 39	280 34	280 29	280 25
11	281 58	281 53	281 47	281 42	281 37	281 32	281 27
12	283 03	282 57	282 51	282 45	282 40	282 34	282 29
13	284 08	284 02	283 55	283 49	283 43	283 37	283 31
14	285 13	285 06	284 59	284 53	284 46	284 40	284 33
15	286 17	286 10	286 03	285 56	285 49	285 42	285 35
16	287 22	287 14	287 07	286 59	286 52	286 45	286 37
17	288 26	288 18	288 11	288 02	287 55	287 47	287 39
18	290 31	289 22	289 14	289 05	288 57	288 49	288 41
19	290 35	290 26	290 17	290 08	290 00	289 51	289 43
20	291 39	291 30	291 20	291 11	291 02	290 53	290 44
21	292 43	292 33	292 23	292 13	292 04	291 55	291 45
22	293 47	293 37	293 26	293 16	293 06	292 56	292 47
23	294 51	294 40	294 29	294 19	294 08	293 58	293 48
24	295 54	295 43	295 32	295 21	295 10	294 59	294 49
25	296 57	296 46	296 35	296 23	296 12	296 01	295 50
26	298 01	297 49	297 37	297 25	297 14	297 02	296 51
27	299 04	298 51	298 39	298 27	298 15	298 03	297 51
28	300 07	299 54	299 41	299 28	299 16	299 04	298 52
29	301 09	300 56	300 43	300 30	300 17	300 05	299 53

N. Lats.	*0*	*1*	*2*	*3*	*4*	*5*	*6*
♒	302 12	301 58	301 45	301 31	301 18	301 05	300 53
01	303 14	303 00	302 47	302 33	302 19	302 06	301 53
02	304 16	304 02	303 48	303 34	303 20	303 06	302 53
03	305 18	305 03	304 50	304 35	304 21	304 07	303 53
04	306 20	306 05	305 51	305 36	305 22	305 07	304 53
05	307 22	307 07	306 52	306 36	306 22	306 07	305 52
06	308 24	308 08	307 53	307 37	307 22	307 07	306 52
07	309 25	309 09	308 54	308 37	308 22	308 07	307 51
08	310 26	310 10	309 54	309 37	309 22	309 06	308 50
09	311 27	311 10	310 54	310 37	310 21	310 05	309 49
10	312 28	312 11	311 54	311 37	311 21	311 04	310 48
11	313 28	313 11	312 54	312 37	312 20	312 03	311 47
12	314 29	314 11	313 54	313 37	313 19	313 02	312 46
13	315 29	315 11	314 54	314 36	314 18	314 01	313 44
14	316 29	316 11	315 53	315 35	315 17	315 00	314 43
15	317 29	317 10	316 52	316 34	316 16	315 58	315 41
16	318 29	318 10	317 51	317 33	317 15	316 57	316 39
17	319 29	319 09	318 50	318 32	318 14	317 55	317 37
18	320 28	320 08	319 49	319 30	319 12	318 53	318 35
19	321 27	321 07	320 48	320 29	320 10	319 51	319 33
20	322 26	322 06	321 47	321 27	321 08	320 49	320 31
21	323 25	323 04	322 45	322 25	322 06	321 47	321 28
22	324 23	324 03	323 43	323 23	323 04	322 45	322 25
23	325 22	325 01	324 41	324 21	324 01	323 42	323 22
24	326 20	325 59	325 39	325 19	324 59	324 39	324 19
25	327 18	326 57	328 37	326 17	325 56	325 36	325 16
26	328 16	327 55	327 35	327 14	326 53	326 33	326 13
27	329 14	328 53	328 32	328 11	327 50	327 30	327 10
28	330 11	329 51	329 29	329 08	328 47	928 27	328 06
29	331 09	330 47	330 26	330 05	329 44	329 25	329 03

N. Lats.	*0*	*1*	*2*	*3*	*4*	*5*	*6*
♓	332 06	331 44	331 23	331 02	330 41	330 20	329 59
01	333 04	332 41	332 20	331 59	331 38	331 16	330 55
02	334 01	333 38	333 17	332 55	332 34	332 12	331 51
03	334 58	334 35	334 13	333 51	333 30	333 08	332 47
04	335 55	335 32	335 10	334 48	334 26	334 04	333 43
05	336 51	336 29	336 07	335 44	335 22	335 00	334 39
06	337 48	337 25	337 03	336 40	336 18	335 56	335 34
07	338 44	338 22	337 59	337 36	337 14	336 52	336 30
08	339 40	339 18	338 55	338 32	338 10	337 48	337 26
09	340 37	340 14	339 51	339 28	339 06	338 43	338 21
10	341 33	341 10	340 47	340 24	340 02	339 39	339 17
11	342 29	342 06	341 43	341 20	340 58	340 35	340 12
12	343 25	343 02	342 39	342 16	341 53	341 30	341 07
13	344 20	343 58	343 35	343 12	342 49	342 25	342 02
14	345 16	344 53	344 30	344 07	343 44	343 20	342 57
15	346 12	345 48	345 25	345 02	344 39	344 15	343 52
16	347 07	346 44	346 21	345 57	345 34	345 10	344 47
17	348 03	347 40	347 17	346 52	346 29	346 05	345 42
18	348 58	348 35	348 12	347 47	347 24	347 00	346 37
19	349 54	349 31	349 07	348 43	348 19	347 55	347 32
20	350 49	350 26	350 03	349 38	349 14	348 50	348 27
21	351 44	351 21	350 57	350 33	350 09	349 45	349 22
22	352 39	352 16	351 52	351 28	351 04	350 40	350 17
23	353 35	353 11	352 47	352 23	351 59	351 35	351 12
24	354 30	354 06	353 42	353 18	352 54	352 30	352 07
25	355 25	355 01	354 38	354 14	353 50	353 26	353 02
26	356 20	355 57	355 33	355 09	354 45	354 21	353 57
27	357 15	356 52	356 28	356 04	355 40	355 16	354 52
28	358 10	357 47	357 23	356 59	356 35	356 11	355 47
29	359 05	358 42	358 18	357 54	357 30	357 00	355 42

S. Lats.	*0*	*1*	*2*	*3*	*4*	*5*	*6*
♈	00 00	00 23	00 47	01 11	01 35	01 59	02 23
01	00 55	01 18	01 42	02 06	02 30	02 54	03 18
02	01 50	02 13	02 37	03 01	03 25	03 49	04 13
03	02 45	03 08	03 32	03 56	04 20	04 44	05 08
04	03 40	04 03	04 27	04 51	05 15	05 39	06 03
05	04 35	04 58	05 22	05 46	06 10	06 34	06 58
06	05 30	05 54	06 18	06 42	07 06	07 30	07 53
07	06 25	06 49	07 13	07 37	08 01	08 25	08 48
08	07 21	07 44	08 08	08 32	08 56	09 20	09 43
09	08 16	08 40	09 04	09 28	09 51	10 15	10 38
10	09 11	09 35	09 59	10 23	10 46	11 10	11 33
11	10 06	10 30	10 54	11 18	11 41	12 05	12 28
12	11 02	11 25	11 49	12 13	12 36	13 00	13 23
13	11 57	12 20	12 44	13 08	13 31	13 55	14 18
14	12 53	13 16	13 39	14 03	14 26	14 50	15 13
15	13 48	14 12	14 35	14 58	15 21	15 45	16 08
16	14 44	15 07	15 30	15 53	16 16	16 40	17 03
17	15 40	16 02	16 25	16 48	17 11	17 35	17 58
18	16 35	16 58	17 21	17 44	18 07	18 30	18 53
19	17 31	17 54	18 17	18 40	19 02	19 25	19 48
20	18 27	18 50	19 13	19 36	19 58	20 21	20 43
21	19 23	19 46	20 09	20 23	20 54	21 17	21 39
22	20 20	20 42	21 05	21 28	21 50	22 12	22 34
23	21 16	21 38	22 01	22 24	22 46	23 08	23 30
24	22 12	22 35	22 57	23 20	23 42	24 04	24 26
25	23 09	23 31	23 53	24 16	24 38	25 00	25 21
26	24 06	24 28	24 50	25 12	25 34	25 55	26 17
27	25 02	25 25	25 47	26 09	26 30	26 52	27 13
28	25 59	26 22	26 43	27 05	27 26	27 48	28 09
29	26 57	27 19	27 40	28 01	28 22	28 44	29 05

S. Lats.	*0*	*1*	*2*	*3*	*4*	*5*	*6*
♉	27 54	28 16	28 37	28 58	29 19	29 40	30 01
01	28 51	29 13	29 34	29 55	30 16	30 37	30 57
02	29 49	30 10	30 31	30 52	31 13	31 34	31 54
03	30 46	31 07	31 28	31 49	32 10	32 31	32 51
04	31 44	32 05	32 25	32 46	33 07	33 27	33 47
05	32 42	33 03	33 23	33 43	34 04	34 24	34 44
06	33 40	34 01	34 25	34 41	35 01	35 21	35 41
07	34 38	34 59	35 19	35 39	35 58	36 18	36 38
08	35 37	35 57	36 17	36 37	36 56	37 15	37 35
09	36 36	36 56	37 15	37 35	37 54	38 13	38 32
10	37 34	37 54	38 13	38 33	38 52	39 11	39 29
11	38 33	38 53	39 12	39 31	39 50	40 09	40 27
12	39 33	39 52	40 11	40 30	40 48	41 07	41 25
13	40 32	40 51	41 10	41 28	41 46	42 05	42 23
14	41 31	41 50	42 09	42 27	42 45	43 03	43 21
15	42 31	42 50	43 08	43 26	43 44	44 02	44 19
16	43 31	43 49	44 07	44 25	44 43	45 00	45 17
17	44 31	44 49	45 06	45 24	45 42	45 59	46 15
18	45 31	45 49	46 06	46 23	46 41	46 58	47 14
19	46 32	46 49	47 06	47 23	47 40	47 57	48 13
20	47 32	47 49	48 06	48 23	48 39	48 56	49 12
21	48 33	48 50	49 06	49 23	49 39	49 55	50 11
22	49 34	49 50	50 06	50 23	50 38	50 54	51 10
23	50 35	50 51	51 06	51 23	51 38	51 53	52 09
24	51 36	51 52	52 07	52 23	52 38	52 53	53 08
25	52 38	52 53	53 08	53 24	53 38	53 53	54 08
26	53 40	53 55	54 09	54 24	54 38	54 53	55 07
27	54 42	54 56	55 11	55 25	55 39	55 53	56 07
28	55 44	55 58	56 12	56 26	56 40	56 54	57 07
29	56 46	57 00	57 13	57 27	57 41	57 54	58 07

S. Lats.	0	1	2	3	4	5	6
♊	57 48	58 02	58 15	58 29	58 42	58 55	59 07
01	58 51	59 04	59 17	59 30	59 43	59 55	60 07
02	59 53	60 06	60 19	60 31	60 44	60 56	61 08
03	60 56	61 08	61 21	61 33	61 46	61 57	62 09
04	61 59	62 11	62 23	62 35	62 48	62 58	63 09
05	63 03	63 14	63 25	63 37	63 50	63 59	64 10
06	64 06	64 17	64 28	64 39	64 52	65 01	65 11
07	65 09	65 20	65 31	65 41	65 54	66 02	66 12
08	66 13	66 23	66 34	66 44	66 56	67 04	67 13
09	67 17	67 27	67 37	67 46	67 58	68 06	68 15
10	68 21	68 30	68 40	68 49	68 59	69 07	69 16
11	69 25	69 34	69 43	69 52	70 01	70 09	70 17
12	70 29	70 38	70 46	70 55	71 03	71 11	71 19
13	71 34	71 42	71 49	71 58	72 05	72 13	72 21
14	72 38	72 46	72 53	73 01	73 08	73 15	73 23
15	73 43	73 50	73 57	74 04	74 11	74 18	74 25
16	74 47	74 54	75 01	75 07	75 14	75 20	75 27
17	75 52	75 58	76 05	76 11	76 17	76 22	76 29
18	76 57	77 03	77 09	77 15	77 20	77 25	77 31
19	78 02	78 07	78 13	78 18	78 23	78 28	78 33
20	79 07	79 12	79 17	79 21	79 26	79 31	79 35
21	80 12	80 17	80 21	80 25	80 29	80 34	80 38
22	81 17	81 21	81 25	81 28	81 32	81 36	81 40
23	82 22	82 25	82 29	82 32	82 35	82 39	82 42
24	83 28	83 30	83 33	83 36	83 39	83 42	83 45
25	84 33	84 35	84 37	84 40	84 42	84 45	84 47
26	85 38	85 40	85 41	85 44	85 45	85 48	85 49
27	86 44	86 45	86 46	86 48	86 49	86 51	86 52
28	87 49	87 50	87 50	87 52	87 52	87 54	87 54
29	88 55	88 55	88 55	88 56	88 56	88 57	88 57

S. Lats.	*0*	*1*	*2*	*3*	*4*	*5*	*6*
♋	90 00	90 00	90 00	90 00	90 00	90 00	90 00
01	91 05	91 05	91 05	91 04	91 04	91 03	91 03
02	92 11	92 10	92 10	92 08	92 08	92 06	92 06
03	93 16	93 15	93 14	93 12	93 11	93 09	93 08
04	94 22	94 20	94 19	94 16	94 15	94 12	94 11
05	95 27	95 25	95 23	95 20	95 18	95 15	95 13
06	96 32	96 30	96 27	96 24	96 21	96 18	96 15
07	97 38	97 35	97 31	97 28	97 25	97 21	97 18
08	98 43	98 39	98 35	98 32	98 28	98 24	98 20
09	99 48	99 43	99 39	99 35	99 31	99 26	99 22
10	100 53	100 48	100 43	100 39	100 34	100 29	100 25
11	101 58	101 53	101 47	101 42	101 37	101 32	101 27
12	103 03	102 57	102 51	102 45	102 40	102 34	102 29
13	104 08	104 02	103 55	103 49	103 43	103 37	103 31
14	105 13	105 06	104 59	104 52	104 46	104 40	104 33
15	106 17	106 10	106 03	105 56	105 49	105 42	105 35
16	107 22	107 14	107 07	106 59	106 52	106 45	106 37
17	108 26	108 18	108 11	108 02	107 55	107 47	107 39
18	109 31	109 22	109 14	109 05	108 57	108 49	108 41
19	110 35	110 26	110 17	110 08	110 00	109 51	109 43
20	111 39	111 30	111 20	111 11	111 02	110 53	110 44
21	112 43	112 33	112 23	112 13	112 04	111 54	111 45
22	113 47	113 37	113 26	113 16	113 06	112 56	112 47
23	114 51	114 40	114 29	114 19	114 08	113 58	113 48
24	115 54	115 43	115 32	115 21	115 10	114 59	114 49
25	116 57	116 46	116 35	116 23	116 12	116 01	115 50
26	118 01	117 49	117 37	117 25	117 14	117 02	116 51
27	119 04	118 51	118 39	118 27	118 15	118 03	117 52
28	120 07	119 54	119 41	119 29	119 16	119 04	118 52
29	121 09	120 56	120 43	120 30	120 17	120 05	119 53

S. Lats.	*0*	*1*	*2*	*3*	*4*	*5*	*6*
♌	122 12	121 58	121 45	121 31	121 18	121 5	120 53
01	123 14	123 00	122 47	122 33	122 19	122 06	121 53
02	124 16	124 02	123 48	123 34	123 20	123 06	122 53
03	125 18	125 03	124 49	124 35	124 21	124 07	123 53
04	126 20	126 05	125 51	125 36	125 22	125 07	124 53
05	127 22	127 07	126 52	126 36	126 22	126 07	125 52
06	128 24	128 08	127 53	127 37	127 22	127 07	126 52
07	129 25	129 09	128 54	128 37	128 22	128 07	127 51
08	130 26	130 10	129 54	129 37	129 22	129 06	128 50
09	131 27	131 10	130 54	130 37	130 21	130 05	129 49
10	132 28	132 11	131 54	131 37	131 21	131 04	130 48
11	133 28	133 11	132 54	132 37	132 20	132 03	131 47
12	134 29	134 11	133 54	133 37	133 19	133 02	132 46
13	135 29	135 11	134 54	134 36	134 18	134 01	133 45
14	136 29	136 11	135 53	135 35	135 17	135 00	134 43
15	137 29	137 10	136 52	136 34	136 16	135 58	135 41
16	138 29	138 10	137 51	137 33	137 15	136 57	136 39
17	139 28	139 09	138 50	138 32	138 14	137 55	137 37
18	140 28	140 08	139 49	139 30	139 13	138 53	138 35
19	141 27	141 07	140 48	140 29	140 10	139 51	139 33
20	142 26	142 06	141 47	141 27	141 08	140 49	140 31
21	143 25	143 04	142 45	142 25	142 06	141 47	141 28
22	144 23	144 03	143 43	143 23	143 04	142 45	142 25
23	145 22	145 01	144 41	144 21	144 02	143 42	143 22
24	146 20	145 59	145 39	145 19	144 59	144 39	144 19
25	147 18	146 57	146 37	146 17	145 56	145 36	145 16
26	148 16	147 55	147 35	147 14	146 53	146 33	146 13
27	149 14	148 53	148 32	148 11	147 50	147 29	147 09
28	150 11	149 50	149 29	149 08	148 47	148 26	148 06
29	151 09	150 47	150 26	150 05	149 44	149 25	149 03

S. Lats.	*0*	*1*	*2*	*3*	*4*	*5*	*6*
♍	152 06	151 44	151 23	151 02	150 41	150 20	149 59
01	153 04	152 41	152 20	151 59	151 38	151 16	150 55
02	154 01	153 38	153 17	152 55	152 34	152 12	151 51
03	154 58	154 35	154 13	153 51	153 30	153 08	152 47
04	155 54	155 32	155 10	154 58	154 26	154 04	153 43
05	156 51	156 29	156 07	155 54	155 22	155 00	154 39
06	157 48	157 25	157 03	156 40	156 18	155 56	155 34
07	158 44	158 22	157 59	157 36	157 14	156 52	156 30
08	159 40	159 18	158 55	158 32	158 10	157 48	157 26
09	160 37	160 14	159 51	159 28	159 06	158 43	158 21
10	161 33	161 10	160 47	160 24	160 02	159 39	159 17
11	162 29	162 06	161 43	161 20	160 58	160 35	160 12
12	163 25	163 02	162 39	162 16	161 53	161 30	161 07
13	164 20	163 58	163 35	163 12	162 40	162 25	162 02
14	165 16	164 53	164 30	164 07	163 44	163 20	162 57
15	166 12	165 48	165 25	165 02	164 39	164 15	163 52
16	167 07	166 44	166 21	165 57	165 34	165 10	164 47
17	168 03	167 40	167 17	166 52	166 29	166 05	165 42
18	168 58	168 35	168 12	167 47	167 24	167 00	166 37
19	169 54	169 31	169 07	168 43	168 19	167 55	167 32
20	170 49	170 26	170 02	169 38	169 14	168 50	168 27
21	171 44	171 21	170 57	170 33	170 09	169 45	169 22
22	172 39	172 16	171 52	171 28	171 04	170 40	170 17
23	173 35	173 11	172 47	172 23	171 59	171 35	171 12
24	174 30	174 06	173 42	173 18	172 54	172 30	172 07
25	175 25	175 02	174 38	174 14	173 50	173 26	173 02
26	176 20	175 57	175 33	175 09	174 45	174 21	173 57
27	177 15	176 52	176 28	176 04	175 40	175 16	174 52
28	178 10	177 47	177 23	176 59	176 35	176 11	175 47
29	179 05	178 42	178 18	177 54	177 30	177 06	176 42

S. Lats.	*0*	*1*	*2*	*3*	*4*	*5*	*6*
♎	180 00	179 37	179 13	178 49	178 25	178 01	177 37
01	180 55	180 32	180 08	179 44	179 20	178 56	178 32
02	181 50	181 27	181 03	180 39	180 15	179 51	179 27
03	182 45	182 22	181 58	181 34	181 10	180 46	180 22
04	183 40	183 17	182 53	182 29	182 05	181 41	181 17
05	184 35	184 12	183 48	183 24	183 00	182 36	182 12
06	185 30	185 07	184 43	184 19	183 55	183 31	183 07
07	186 25	186 02	185 38	185 14	184 50	184 26	184 02
08	187 21	186 57	186 33	186 09	185 45	185 21	184 57
09	188 16	187 52	187 28	187 04	186 40	186 16	185 52
10	189 11	188 47	188 23	187 59	187 35	187 11	186 47
11	190 06	189 42	189 11	188 55	188 31	188 07	187 43
12	191 02	190 38	190 14	189 51	189 27	189 03	188 39
13	191 57	191 33	191 09	190 46	190 22	189 58	189 34
14	192 53	192 29	192 05	191 42	191 18	190 54	190 30
15	193 48	193 25	193 01	192 38	192 14	191 50	191 26
16	194 44	194 20	193 57	193 34	193 10	192 46	192 22
17	195 40	195 16	194 53	194 30	194 06	193 42	193 18
18	196 35	196 12	195 49	195 26	195 02	194 39	194 15
19	197 31	197 08	196 45	196 22	195 58	195 35	195 11
20	198 27	198 04	197 41	197 18	196 54	196 31	196 07
21	199 23	199 00	198 37	198 14	197 51	197 28	197 04
22	200 20	199 56	199 33	199 11	198 48	198 25	198 01
23	201 16	200 53	200 30	200 08	199 45	199 22	198 58
24	202 12	201 50	201 27	201 05	200 42	200 19	199 55
25	203 09	202 47	202 24	202 02	201 39	201 16	200 52
26	204 06	203 44	203 21	202 59	202 36	202 13	201 50
27	205 02	204 41	204 19	203 57	203 34	203 11	202 48
28	205 59	205 38	205 16	204 54	204 31	204 09	203 46
29	206 57	206 35	206 13	205 51	205 29	205 07	204 44

S. Lats.	*0*	*1*	*2*	*3*	*4*	*5*	*6*
♏	207 54	207 33	207 11	206 49	206 27	206 05	205 43
01	208 51	208 30	208 08	207 47	207 25	207 03	206 40
02	209 49	209 27	209 06	208 45	208 23	208 01	207 38
03	210 46	210 25	210 04	209 43	209 21	208 59	208 37
04	211 44	211 23	211 02	210 41	210 18	209 58	209 36
05	212 42	212 21	212 00	211 39	211 19	210 57	210 35
06	213 40	213 20	212 59	212 38	212 17	211 56	211 34
07	214 38	214 18	213 58	213 37	213 16	212 55	212 33
08	215 37	215 17	214 57	214 36	214 15	213 54	213 33
09	216 36	216 16	215 56	215 36	215 15	214 54	214 33
10	217 34	217 15	216 55	216 35	216 15	215 54	215 33
11	218 33	218 14	217 55	217 35	217 15	216 54	216 33
12	219 33	219 14	218 54	218 35	218 15	217 55	217 34
13	220 32	220 13	219 54	219 35	219 15	218 56	218 35
14	221 31	221 13	220 54	220 35	220 16	219 57	219 36
15	222 31	222 13	221 54	221 36	221 17	220 58	220 38
16	223 31	223 13	222 54	222 36	222 18	221 59	221 39
17	224 31	224 13	223 55	223 37	223 19	223 00	222 40
18	225 31	225 14	224 56	224 38	224 20	224 01	223 42
19	226 32	226 14	225 57	225 39	225 21	225 03	224 44
20	227 32	227 15	226 58	226 40	226 23	226 06	225 46
21	228 33	228 16	227 59	227 42	227 25	227 07	226 49
22	229 34	229 17	229 00	228 44	228 27	228 09	227 52
23	230 35	230 18	230 02	229 46	229 29	229 12	228 55
24	231 36	231 20	231 04	230 48	230 32	230 15	229 58
25	232 38	232 22	232 06	231 51	231 35	231 18	231 02
26	233 40	233 24	233 09	232 54	232 38	232 22	232 06
27	234 42	234 27	234 12	233 57	233 42	233 26	233 10
28	235 44	235 29	235 15	235 00	234 45	234 30	234 14
29	236 46	236 32	236 18	236 03	235 49	235 34	235 18

S. Lats.	*0*	*1*	*2*	*3*	*4*	*5*	*6*
♐	237 48	237 35	237 21	237 07	236 53	236 38	236 23
01	238 51	238 38	238 24	238 10	237 57	237 42	237 28
02	239 53	239 41	239 28	239 14	239 01	238 47	238 33
03	240 56	240 44	240 31	240 18	240 05	239 52	239 38
04	241 59	241 47	241 35	241 22	241 10	240 57	240 44
05	243 03	242 51	242 39	242 27	242 15	242 02	241 50
06	244 06	243 55	243 43	243 32	243 20	243 08	242 56
07	245 09	244 59	244 47	244 37	244 25	244 13	244 02
08	246 13	246 03	245 52	245 42	245 30	245 19	245 08
09	247 17	247 07	246 57	246 47	246 36	246 25	246 14
10	248 21	248 11	248 12	247 52	247 42	247 31	247 21
11	249 25	249 16	249 07	248 57	248 48	248 38	248 28
12	250 29	250 21	250 12	250 03	249 54	249 45	249 35
13	251 34	251 26	251 17	251 09	251 00	250 51	250 42
14	252 38	252 31	252 22	252 15	252 06	251 58	251 49
15	253 43	253 36	253 28	253 21	253 13	253 05	252 57
16	254 47	254 41	254 33	254 27	254 19	254 12	254 04
17	255 52	255 46	255 39	255 33	255 26	255 19	255 12
18	256 57	256 51	256 45	256 39	256 33	256 27	256 20
19	258 02	257 56	257 51	257 45	257 40	257 34	257 28
20	259 07	259 02	258 57	258 52	258 47	258 41	258 36
21	260 12	260 08	260 03	259 59	259 54	259 49	259 44
22	261 17	261 13	261 09	261 05	261 01	260 56	260 52
23	262 22	262 18	262 15	262 11	262 08	262 04	262 00
24	263 28	263 24	263 21	263 18	263 15	263 12	263 09
25	264 33	264 30	264 27	264 25	264 22	264 20	264 17
26	265 38	265 36	265 33	265 32	265 29	265 28	265 26
27	266 44	266 42	266 40	266 39	266 37	266 36	266 34
28	267 49	267 48	267 46	267 46	267 44	267 44	267 43
29	268 55	268 54	268 53	268 53	268 52	268 52	268 52

S. Lats.	*0*	*1*	*2*	*3*	*4*	*5*	*6*
♑	270 00	270 00	270 00	270 00	270 00	270 00	270 00
01	271 05	271 06	271 07	271 07	271 08	271 08	271 09
02	272 11	272 12	272 14	272 14	272 16	272 16	272 18
03	273 16	273 18	273 20	273 21	273 23	273 24	273 26
04	274 22	274 24	274 26	274 28	274 31	274 32	274 34
05	275 27	275 30	275 33	275 35	275 38	275 40	275 43
06	276 32	276 36	276 39	276 42	276 45	276 48	276 51
07	277 38	277 41	277 45	277 50	277 52	277 56	278 00
08	278 43	278 47	278 51	278 55	278 59	279 04	279 08
09	279 48	279 52	279 57	280 01	280 06	280 11	280 16
10	280 53	280 58	281 03	281 08	281 13	281 19	281 24
11	281 58	282 04	282 09	282 15	282 20	282 26	282 32
12	283 03	283 09	283 15	283 21	283 27	283 33	283 40
13	284 08	284 14	284 21	284 27	284 34	284 41	284 48
14	285 13	285 19	285 27	285 33	285 41	285 48	285 56
15	286 17	286 24	286 32	286 39	286 47	286 55	287 03
16	287 22	287 29	287 38	287 45	287 54	288 02	288 11
17	288 26	288 34	288 43	288 51	289 00	289 09	289 18
18	289 31	289 39	289 48	289 57	290 06	290 15	290 25
19	290 35	290 44	290 53	291 03	291 12	291 22	291 32
20	291 39	291 49	291 58	292 08	292 18	292 29	292 39
21	292 43	292 53	293 03	293 13	293 24	293 35	293 46
22	293 47	293 57	294 08	294 18	294 30	294 41	294 52
23	294 51	295 01	295 13	295 23	295 35	295 47	295 58
24	295 54	296 05	296 17	296 28	296 40	296 53	297 04
25	296 57	297 09	297 21	297 33	297 45	297 58	298 10
26	298 01	298 13	298 25	298 38	298 50	299 03	299 16
27	299 04	299 16	299 29	299 41	299 55	300 08	300 22
28	300 07	300 19	300 33	300 46	300 59	301 13	301 27
29	301 09	301 22	301 36	301 50	302 03	302 18	302 32

S. Lats.	*0*	*1*	*2*	*3*	*4*	*5*	*6*
♒	302 12	302 25	302 39	302 53	303 07	303 22	303 37
01	303 14	303 28	303 42	303 57	304 11	304 26	304 42
02	304 16	304 31	304 45	305 00	305 15	305 30	305 46
03	305 18	305 33	305 48	306 03	306 18	306 34	306 50
04	306 20	306 36	306 51	307 06	307 22	307 38	307 54
05	307 22	307 38	307 54	308 09	308 25	308 42	308 58
06	308 24	308 40	308 56	309 12	309 28	309 44	310 02
07	309 25	309 42	309 58	310 14	310 31	310 48	311 05
08	310 26	310 43	311 00	311 16	311 38	311 51	312 08
09	311 27	311 44	312 01	312 18	312 35	312 53	313 11
10	312 28	312 45	313 02	313 20	313 37	313 55	314 14
11	313 28	313 46	314 03	314 21	314 39	314 57	315 16
12	314 29	314 46	315 04	315 22	315 40	315 59	316 18
13	315 29	315 47	316 05	316 23	316 41	317 00	317 20
14	316 29	316 47	317 06	317 24	317 42	318 01	318 21
15	317 29	317 47	318 06	318 24	318 43	319 02	319 22
16	318 29	318 47	319 06	319 25	319 44	320 03	320 24
17	319 28	319 47	320 06	320 25	320 45	321 04	321 25
18	320 27	320 46	321 06	321 25	321 45	322 05	322 26
19	321 27	321 46	322 06	322 25	322 45	323 06	323 27
20	322 26	322 45	323 05	323 25	323 45	324 06	324 27
21	323 25	323 44	324 04	324 24	324 45	325 06	325 27
22	324 23	324 43	325 03	325 24	325 45	326 06	326 27
23	325 22	325 42	326 02	326 23	326 44	327 05	327 27
24	326 20	326 40	327 01	327 22	327 43	328 04	328 26
25	327 18	327 39	328 00	328 21	328 42	329 03	329 25
26	328 16	328 37	328 58	329 19	329 41	330 02	330 24
27	329 14	329 35	329 56	330 17	330 39	331 01	331 23
28	330 11	330 33	330 54	331 15	331 37	331 59	332 22
29	331 09	331 30	331 52	332 13	332 35	332 57	333 20

S. Lats.	*0*	*1*	*2*	*3*	*4*	*5*	*6*
♓	332 06	332 28	332 49	333 11	333 33	333 55	334 18
01	333 04	333 25	333 47	334 09	334 31	334 53	335 16
02	334 01	334 22	334 44	335 06	335 29	335 51	336 14
03	334 58	335 19	335 41	336 03	336 26	336 49	337 12
04	335 55	336 16	336 39	337 01	337 24	337 47	338 10
05	336 51	337 13	337 36	337 58	338 21	338 44	339 08
06	337 48	338 10	338 33	338 55	339 18	339 41	340 05
07	338 44	339 07	339 30	339 52	340 15	340 38	341 02
08	339 40	340 04	340 27	340 49	341 12	341 35	341 59
09	340 37	341 00	341 23	341 46	342 09	342 32	342 56
10	341 33	341 56	342 19	342 42	343 06	343 29	343 53
11	342 29	342 52	343 15	343 38	344 02	344 25	344 49
12	343 25	343 48	344 11	344 34	344 58	345 21	345 45
13	344 20	344 44	345 07	345 30	345 54	346 18	346 42
14	345 16	345 40	346 03	346 26	346 50	347 14	347 38
15	346 12	346 35	346 59	347 22	347 46	348 10	348 34
16	347 07	347 31	347 55	348 18	348 42	349 06	349 30
17	348 03	348 27	348 51	349 14	349 38	350 02	350 26
18	348 58	349 22	349 46	350 09	350 33	350 57	351 21
19	349 54	350 18	350 42	351 05	351 29	351 53	352 17
20	350 49	351 13	351 37	352 01	352 25	352 49	353 13
21	351 44	352 08	352 32	352 56	353 20	353 44	354 08
22	352 39	353 03	353 27	353 51	354 15	354 39	355 03
23	353 35	353 58	354 22	354 46	355 10	355 34	355 58
24	354 30	354 53	355 17	355 41	356 05	356 29	356 53
25	355 25	355 48	356 12	356 36	357 00	357 24	357 48
26	356 20	356 43	357 07	357 31	357 55	358 19	358 43
27	357 15	357 38	358 02	358 26	358 50	359 14	359 38
28	358 10	358 33	358 57	359 21	359 45	360 09	360 33
29	359 05	359 28	359 52	360 16	360 40	361 04	361 28

www.ingramcontent.com/pod-product-compliance
Lightning Source LLC
LaVergne TN
LVHW020647100826
845148LV00012B/2358

* 9 7 8 0 8 6 6 9 0 6 1 5 9 *